P9-DGC-005

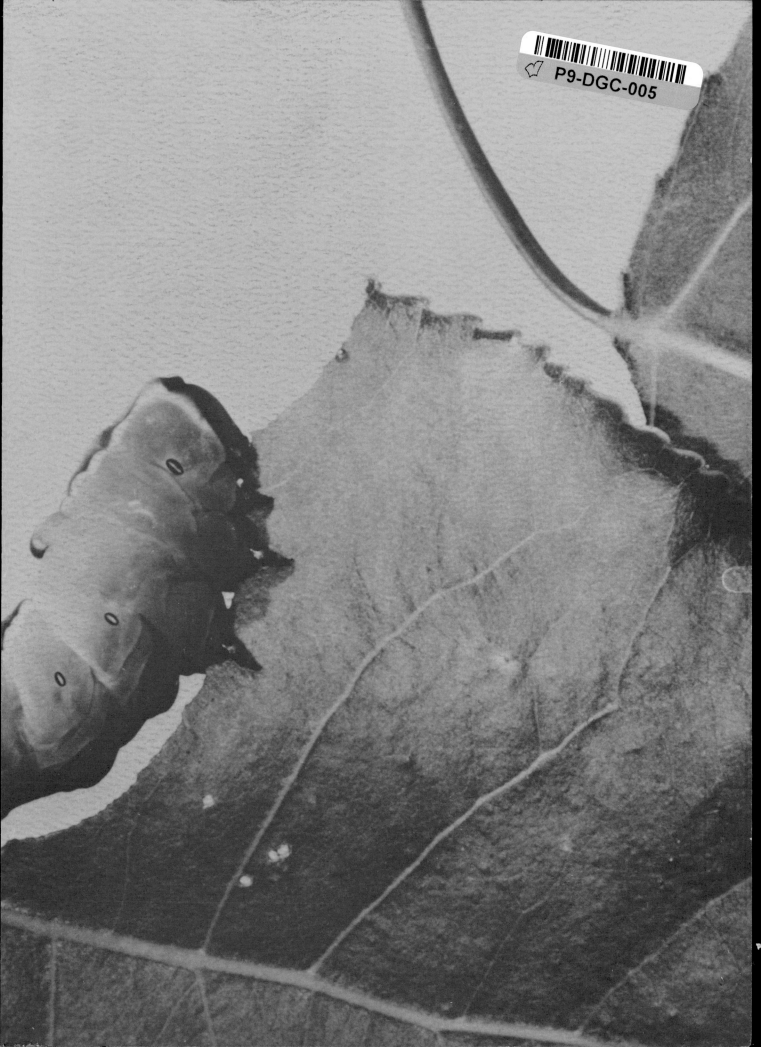

OTHER BOOKS BY ROGER CARAS

Roger Caras

VENOMOUS

OF THE

Prentice-Hall, Inc.

ANIMALS
WORLD

Englewood Cliffs, N. J.

GARDNER WEBB COLLEGE LIBRARY

All photographs in this book, both black and white and full color, were provided by Photo Researchers, Inc.
All line drawings in this book were done by Theodore Xaras.

Venomous Animals of the World by Roger Caras
Copyright © 1974 by Roger A. Caras

Copyright under International and Pan American Copyright Conventions

All rights reserved. No part of this book may be reproduced in any form or by any means, except for the inclusion of brief quotations in a review, without permission in writing from the publisher.

Printed in the United States of America

Prentice-Hall International, Inc., London
Prentice-Hall of Australia, Pty. Ltd., Sydney
Prentice-Hall of Canada, Ltd., Toronto
Prentice-Hall of India Private Ltd., New Delhi
Prentice-Hall of Japan, Inc., Tokyo

10 9 8 7 6 5 4 3 2 1

Library of Congress Cataloging in Publication Data

Caras, Roger A
Venomous animals of the world.
Bibliography: p.
1. Poisonous animals. I. Title.
[DNLM: 1. Animals, Poisonous. 2. Behavior, Animal.
3. Venoms. WD400 C261v]
QL100.C37   591.6'9   74-8633
0-13-941526-2

QL
100
C 37

This book is for Harris Warren Hantman, M.D.,
for life and other favors. . . .

ACKNOWLEDGMENTS

Special gratitude is due to so very many people that listing them here is quite impossible. Their names appear in the Bibliography and their contributions are obvious. The equation is simple and direct: They did the research, I studied their observations, and together we made this book happen.

My very special thanks, though, go to Dr. Sherman Minton of Indianapolis. He read each chapter as it was written, commented on content, and suggested ways of making the story clearer to the general reader. It must be noted, however, that there are many stages between a first technical reading and publication. Any errors that may occur crept in after Dr. Minton had seen the manuscript and are clearly my fault and not his.

Typing a manuscript may seem to the uninitiated to be a simple mechanical task. That is not the case when a first draft is torn up and redigested, with notes written sideways and upside down, scores of arrows indicating transposition, and endless marginal remarks. My thanks to Linda Schwob for wading through the mess and making the manuscript printable.

FOREWORD

This is a unique book that summarizes for the general reader much of what is known about venomous animals and their venoms. A great deal has been written on this morbidly fascinating topic, and Roger Caras has drawn from a formidable bulk of scientific and popular literature as well as from personal contact with authorities throughout the world. To interpret the findings of research scientists and physicians, as well as to evaluate such a great body of anecdote and folklore, requires a special skill. Roger Caras has done this without sacrificing accuracy and he has done it in an interesting and readily understandable manner. This is a book that tells you how a slow-moving cone shell can harpoon a fast-moving fish with a deadly dart, what your chances are of being bitten should you encounter a Gila monster or a coral snake, and where you can find information about the chemistry of tetrodotoxin or samandarine. It presents a multitude of facts that will be useful for settling arguments in classrooms or at cocktail parties, but it also draws attention to the great gaps that remain in man's knowledge of venoms and the animals that produce them.

The power of these animals to cause suffering and death from momentary contact and trivial injury has fascinated and terrified man since paleolithic times. As devil and deity, the snake twists its way through human thought from paintings on cave walls to church services in Appalachia. The scorpion is equally at home in the Bible and the zodiac, and deadly jellyfish are depicted in bark paintings by the aborigines of Arnhem Land. Venomous animals are a part of the world in which man evolved. And they still threaten him as he explores new environments and experiments with novel patterns of behavior. Astronaut Scott Carpenter found the undersea world more hazardous than outer space when he was one of several divers stung by scorpionfish during the Sealab II project off the California coast. A completely unforeseen rattlesnake bite was one of two physical injuries reported by a nude marathon encounter therapy group at Malibu Beach.

The animals that are the subjects of this book are incomparably older than man. The venom apparatus of jellyfish probably has changed little since these animals drifted through Cambrian seas 500 million years ago. Two hundred million years later, the ancestors of the scorpions were the first animals to make the transition from sea to land. Poisonous snakes are a comparatively new development but still have some 30 million years of evolution behind them. Obviously none of these animals became venomous in order to kill people, yet we inevitably think of them in this context.

A recurring theme in this book is "overkill." Why does a spider weighing 1/100 of an ounce *need* a venom strong enough to kill an adult human being? The probable answer is that it doesn't. The whole thing is a biological mistake; a classic example of over-reaction to a stimulus. The comment of Dr. Lewis Thomas, a pathologist at Yale University School of Medicine, regarding the body's reaction to certain bacterial toxins may also apply to many venoms: "When we sense lipopolysaccharide, we are likely to turn on every defense at our disposal; we will bomb, defoliate, blockade, seal off and destroy all the tissues in the area. . . . There is nothing intrinsically poisonous about endotoxin, but it must look awful, or feel awful, when sensed by cells. . . . Sometimes, the mechanisms used for overkill are immunologic, but often . . . they are more primitive kinds of memory. We tear ourselves to pieces because of symbols, and we are more vulnerable to this than to any host of our predators. We are, in effect, at the mercy of our own pentagons most of the time."

Animal venoms are also substances of considerable scientific interest apart from their abilities to cause pain and threaten life. Professor Habermann, Chairman of the Pharmakologisches Institut in Giessen, West Germany, says: "Teleologically, venomous animals practice high-level biochemical pharmacology; they have invented a series of very active, specific, pharmacologically and chemically novel drugs that may be useful in elucidating basic mechanisms of central nervous . . . or membrane . . . functions." Snake venom enzymes and toxins have been used to prepare new compounds for the biochemist, solve problems involving the transmission of nerve impulses or the mechanism of blood clotting, and dissect virus particles. A substance in several snake venoms that stimulates growth of nerves and a factor in cobra venom that selectively destroys certain types of malignant cells hold great potential for future investigation. Venom of the Malay pit viper is the source of an anticoagulant drug widely used in England and under study in the United States. Venoms of invertebrates, especially those of the sea, represent an untapped source of substances potentially important in biology and medicine.

Nearly 22 centuries have passed since Nikandros of Kolophon wrote the first definitive work on plant and animal toxins, yet there remains much to be learned. This must involve workers in many fields. Biochemists and pharmacologists can determine the molecular makeup and site of action of toxins with a degree of precision undreamed of a half century ago, but they need the insights of the taxonomist to identify the animals from whence these toxins came and the observations of the field biologist to determine how they are employed in nature. Ultimately many of the findings of the laboratory must be applied to the treatment of venom poisoning in man where, in the words of one prominent authority, "the mouse makes a very naive human being." Finally there is a need for education, for a great deal of misinformation on poisonous animals is still current. Their dangerous potential has been overrated in some instances, underrated in others. Books such as Roger Caras's *Venomous Animals of the World* help rectify this and are a worthy contribution to the interdisciplinary literature on animal venoms.

Sherman Minton, M.D.
School of Medicine
Department of Microbiology
Indiana University Medical Center

References:
Lewis Thomas, "Notes of a Biology-watcher," *New England Journal of Medicine*, Sept. 14, 1972, pp. 553-555.
E. Habermann, "Bee and Wasp Venoms," *Science*, July 28, 1972, pp. 314-322.

CONTENTS

INTRODUCTION

The subject of this book is the ability of some animals—some thousands of animals, in fact—to inflict a chemical trauma on other organisms. The chemical injury is usually preceded by a small, perhaps even trivial mechanical insult. The real damage is caused by a chemical weapon, a bizarre product of evolution.

We come immediately to the matter of definitions, and a troublesome one it can be. Any reader in natural history knows that clean demarcations are the rarest animals of all. There are mammals that lay eggs, birds that do not fly, fish that walk on land, lizards without legs, frogs that imbed their tadpoles in their own flesh, and anyone who tries to impose a rigid set of guidelines on the natural world is automatically being arbitrary. He is also in for a distorted picture of what things are really like. But, we can attempt some definitions and apologize in advance for the loopholes that must inevitably emerge as we explore our subject.

The words *venomous* and *poisonous*—or *venom* and *poison*, if you wish—are generally used interchangeably. We will make a distinction between them that we must acknowledge will not be acceptable to all. It will make our lives easier, however, at least throughout this book. We shall refer to a *venom* as a substance naturally produced within the body of one animal that can be inflicted on the body of another, usually at will. We will in almost all cases be speaking of a substance that evolved in conjunction with a special device or complex of devices that are used to inflict it under specific circumstances. We will also be speaking of substances that were evolved for this purpose. They came about as survival mechanisms. Their function is to harm other organisms. *Venoms* are not accidents,* *poisons* may be.

*An extreme degree of sensitivity, however, particularly in what would not normally be a target species, may be accidental.

We will take *poison* to mean something that is harmful to ingest. A polar bear's liver can kill a man, but it cannot be said the polar bear evolved a poisonous liver in order to survive predation by man—just about the only creature besides the killer whale (we surmise) that would prey on a polar bear. To make a simple analogy—our point of view in this book will be that a dart dipped in a deadly chemical and fired from a blowpipe is *venomous*, while a mushroom that is deadly to eat is *poisonous*.

Another difference emerges in the distinction we have made. A venom is always produced in special glands, or otherwise ordinary glands especially adapted. This is not true of poisons. Some fish are perfectly safe for man to eat until they in turn eat certain kinds of food that are also safe for man to eat. The combination of fish and fish food, however, becomes dangerous, and several forms of fish poisoning are potentially lethal to trusting human beings. Poisons do not necessarily appear in nature in order to harm us. Some just happen to have a harmful effect. (Such an observation, of course, reflects man's egocentric way of viewing anything that is potentially dangerous to him.)

What about the word *toxin*? That is a little more complex. It has been proposed that the word be used only to designate substances that are either *proteinaceous (nitrogenous organic compounds of high molecular weight synthesized by both plants and animals that yield amino acids under the attack of enzymes)* or *antigenic (a substance produced by an organism that will stimulate the production of antibodies)* or both. This definition is rather too limited for our generalized discussion so we will follow a suggestion made by Dr. W. Vogt (1970) of the Max-Planck Institute for Experimental Medicine in Göttingen, West Germany. He proposed that a toxin fit three basic requirements:

(1) *It should be a naturally occurring substance produced by plants, animals, bacteria.*

(2) *It should be foreign to the victim, or at least be foreign in the quantities received during the act of envenomation.* An example is histamine. It occurs in the human body (and, as we shall be seeing, can cause pain when released), but becomes a toxic substance when injected in larger doses, such as in certain bee stings.

(3) *It should be essentially inimical to the well-being of the organism upon which it is inflicted, at least in the context or dosage under discussion.* An example here is cobra venom. In low doses, under regulated circumstances, it is used in the control of intractable pain and is a merciful medicine. Injected by a frightened or enraged cobra, however, it is a toxin.

Other words we may encounter are *biotoxin* and *antitoxin*. *Biotoxin* is sometimes used synonymously with *venom*, but some investigators (Witkop, 1965) hold that the use should be restricted to proteins only. *Antitoxins* are substances produced by the body to defend itself against the intrusion of toxins elaborated during the metabolism of living organisms like bacteria, for example. The materials manufactured today to combat many venoms are *not antivenoms*, as is so often stated, but *antivenins*.* (A *venin* is a toxic constituent of a venom.)

(As a side note, the word venom is derived from or related to Venus, the goddess of love. It reflects the ability of the substance to *intoxicate*—and envenomation is often referred to as intoxication. The Greek word *tox*—meaning poison—may have derived from *taxus*, the word for yew. Yew wood was used to make bows *(toxon)* which shot *toxicon* or *pharmakon*, both words meaning poison, but the latter also meaning drug. *Ia* in Greek means arrow and in certain combined forms again comes back to medicine and drugs; *iatros* means physician.)

Because the subject is so extraordinarily complex and belongs in the province of the specialist, we shall deal with little chemistry in the chapters that follow. All poisonous and venomous substances fall into two broad categories—protein and nonprotein. The protein substances are by a very wide margin the more difficult to describe or analyze. They are also the more dangerous in almost every case. They are all but impossible to reconstruct in the laboratory. We shall make reference to the chemistry of animal venoms, but little more than that. In passing, we might note that there is not a single venom that has fully yielded to analysis. (Some, like bee venoms, however, are fairly well understood.) Even with the advanced organic and biochemical devices and systems available, most venoms have still managed to retain many of their secrets. To the almost inevitable argument that we are offering a book about chemicals that avoids the discussion of chemistry, we could counter with the equally superficial but perhaps pleasing defense that one can discuss military tactics without delving into the psychoanalytic character of each individual foot soldier. We will discuss venoms in terms of where they are found, how they act, and what problems they pose from the point of view of public health. We shall also speculate from time to time on how they came to be and what their evolutionary impetus may have been. It is just that their precise chemistry, insofar as it has been uncovered to date, would require more background than is fair to assume.

---

*The difference between an *antitoxin* and *antivenin* is more one of semantics than immunology. Their biological nature and mode of action are much the same. *Antivenin* is an Americanization of *antivenene*, a term coined in Edinburgh by Dr. T. R. Fraser.

A subject that shall appear again and again in the pages that follow is *toxicity*, the degree to which an animal is toxic or venomous. I have never heard a reasonable guide or rule that could help correlate toxicity with need as we understand the two phenomena. Why, for example, does the black widow spider have a venom very, very much more toxic than any other spider? Why does a Gila monster that apparently lives on infant rodents and fledgling birds have a venom that can kill, and has killed, 200-pound men? It could be accidental that the venom of a given animal is so seriously toxic to man, yet the degree of toxicity is generally reflected in laboratory experiments on other animals, mammalian and otherwise. While it is true that tolerances to venoms vary, it is also true that a powerful toxin, when used against a reasonably broad cross section of animals, is still a powerful toxin. For me, at least, the level of toxicity, the often extreme potency, remains a profound mystery. If we could unravel that mystery, I think we should know a great deal more not only about venoms but also about the relationship of animals to their environment. We must never lose sight of the fact that although the bizarre powers of venoms tend to make us see them as things or substances apart, they are not that at all. Venoms are a means of coping with the environment. Venoms are survival devices. They are a means for obtaining food or a means to keep from becoming food.*

A second subject that will inevitably occur is the matter of purpose—What is the specific purpose of a given venom? They do not just occur; they have evolved because the tendency in that direction demonstrated survival advantages. The subject is complex and in most instances barely more than speculative. The snakes, the centipedes, the spiders, the shrews, and the two venomous lizards have their venom production apparatus and their venom introduction devices in or around the mouth. It is believed, logically, that their venom-producing glands are modified salivary glands. It is known that the teeth that have become specialized for obtaining maximum efficiency from the venoms are modified from teeth with simpler purposes.** It is not, therefore, unreasonable to assume that the venom of these animals is essentially food-getting. On the other hand, a rattlesnake will bite a man from whom it could never expect to make a meal. The

---

*The role of "accidental" susceptibility cannot, however, be brushed aside. A venom evolved to stop the flight of a passerine bird just might happen to be dangerous to a giraffe. All of this still remains in the area of speculation and theory.

**Once again, we must admit, we are in a speculative area. Recent evidence has shown that snakes whose venom glands have been tied off can still digest prey. That might cast doubt on the theory that digestive enzymes give rise to venoms—then again it might not.

use of the venom in this case is clearly defensive. We may assume a secondary role. What, though, was the *principal* evolutionary impetus? The need to intoxicate prey, the need to drive away large and dangerous animals that were not prey, each equally, or both in some strange combination we do not begin to understand. We must learn a great deal more about animals and how they live, and, most importantly, *how they once lived* before we can answer such questions. Could a major impetus for the development of snake venom have been supplied by animals or influences that no longer exist and whose role in the life of snakes we do not even suspect? Venoms, as we shall see, are an ancient answer to what was perforce an ancient environment. How has it changed—from the venomous animal's point of view?

Some animals (bees) die when they use their venom and that, too, as we shall see, is a mystery. Some animals (hymenoptera and scorpions) have their stinging devices located far from their mouths. Are we to assume that since the venom and the envenomating apparatus did not evolve as part of the ingestive and digestive processes that they are not important in food-getting? Field observations, as we shall see, do not bear out such assumptions.

A further mystery lies in the sexual dimorphism of envenomating apparatuses, or at least venom toxicity. The male platypus is venomous, the female is not. The females of most (*but not all*) spiders are more toxic than the males, or at least so one theory has it. The difference could be in the amount of venom given in a bite. Females need more food since they produce eggs, and they may therefore need more venom. There is no known difference in toxicity between the sexes in snakes. Male hymenoptera cannot sting at all; only females have this power. What all this means is again speculative. Perhaps these facts mean many things.

Among the venomous species are actively venomous animals like snakes and scorpions, and passively venomous ones like caterpillars. The difference between these groups is that the former possess a device for introducing venom *willfully*. The latter must be bumped or brushed to release their venom. In between are jellyfish which when bumped or brushed let fly a veritable shower of venomous darts of microscopic size. Whether this reaction is all involuntary in the jellyfish or at least partially under the control of some strange and primitive central nervous system is not clear. (We assume *primitive* as opposed to *extraordinarily advanced*.)

All of these matters we will consider as we discuss the different groups of animals. We will openly acknowledge that we are frequently puzzled—and the reasons why that is so will be given.

Man has never lived at a time when there were not venomous animals. It is certainly safe to say that all present groups of venomous animals predate man by many millions of years. It is not strange or unusual, therefore, that man has wondered and worried about these animals, and that he has created gods, goddesses, ceremonies, and beliefs to handle a subject which surely was even more troublesome in the past than it is now.

Venoms are generally *but not exclusively* rural problems. While it is true that cobras have bred in a public library in Bombay and black widow spiders are found in New York City, the closer people are to the soil, so to speak, the stronger the likelihood that they will encounter a venomous animal. In ancient times, when the world population was perhaps 99 percent rural, the problem must have been very much more intense than it is today. It is still intense, of course, for many rural populations and the transient rural-urbanized people who retreat to the soil periodically to get their souls recharged.

Venomous snakes figured in the worship of the demigod Imhotep in Egypt. They figured as well in the adoration of Aeklepios in Greece. Throughout the ancient world, as far as we know it, and deep into contemporary indigenous cults and religions, we find the propitiation of venomous animals through adoration. Medicines to combat the effects of venoms go back at least as far as the ancient Greeks and Romans and have been under development and improvement ever since. The actual efficacy of many of the so-called primitive treatments can not really be evaluated. Our traditional Western attitude that our medicine and our medicine alone has clinical validity is a little shakier now than it was even a few years ago. Our view of tribal medicine has been linked historically to our perverted view of tribal man. With the slow collapse of our attitude toward *inferior people* must go our antique attitude toward *inferior medicine*. We do not know how much so-called primitive peoples know about the treatment of venomous injuries. On the surface of things we know rather more. The truth, beneath the surface, may be quite different. I have personally visited areas where European-trained physicians *with antivenins at their disposal* routinely refer snake-bite cases to native practitioners who have never so much as strolled a campus grounds. As will be pointed out, with rather a sense of wonder, recent research into the synergistic relationship of certain viper venoms and ascorbic acid deficiency may shed new light on the oft-encountered use of lime juice in native snake-bite remedies. The fact that laboratory animals respond differently to venoms according to the amount of protein in their diet may also make us look back and move a little less precipitously away from indigenous attitudes toward envenomation.

Venoms and venomous animals represent some of the more puzzling mysteries of the animal kingdom. We have found the doors to some mysteries and we are constantly seeking for others. But each new door we open seems to welcome us to dozens of new ones. As we penetrate this secret world of animal power we are likely to find ourselves being led off in ever new directions. Animal venoms are as potentially useful and important as they are now dangerous and confusing. Some few are already in use in medicine, more almost certainly will be in the years ahead. As we review the profound effects these substances can have on organisms as large and complex as man, we are constantly reminded that anything that can do so much damage must possess keys to the body's basic systems and body chemistry. If that same power could be rechanneled, the gains that could be made in research and in therapy can scarcely be imagined.

There can be no suggestion that we will have a world without venomous animals in the foreseeable future. With our present technology we could only rid this planet of venomous animals by ridding it of ourselves. Any power that could wipe them out would wipe us out first. Man and venomous animals are here to stay. It behooves us to understand their mysteries with rather more depth than we presently do. Lest this appear academic it should be pointed out that this year as many as 60,000 people may die from encounters with venomous animals. If that many people were killed by an atomic bomb or by the accidental explosion of a nuclear power plant, there would not be a single newspaper in the world that would not cover the topic as front-page news. Yet, such a catastrophe would in all likelihood be a one-time event. (An atomic bomb *purposefully* exploded, of course, might be likely to have dire consequences—but, I think our point can still be made.) Mortality from venoms, on the other hand, is a guaranteed annual tally. The only reason it is not front-page news is that we have been living with it since before we ever achieved our present status as a species. In other words, 60,000 deaths a year from envenomation is *old news.* *

*There is, too, something of a difference in the impact of one big event and 60,000 small ones.

It may seem strange to write a book about the damage venoms can do to man without discussing the ways man can combat the damage. The decision to take this approach, however, was not difficult to make. It is possible to list drugs and methods of treatment but it is not possible to impart the judgment needed to choose between them. An episode of envenomation is often a first-class medical emergency, a time for the finest professional help available, not a time for home-grown experts. The zoo men, I know, the herpetologists and others who are constantly exposed to venomous animals, seek the help of doctors when an emergency arises; they do not treat themselves unless there is no other possible course. Zoo keepers do not plunge antivenin into their veins; they go to the refrigerator, take out antivenin, and take it with them to the hospital in the police car.

First aid again requires careful evaluation. It has been believed for too long that all you have to do is slash the victim open and "suck" the venom out. I think more damage is done in many cases by the slashing than by the intoxication.

This is not a book about how to treat envenomation. That discussion would presuppose that both the writer and the reader had had a medical background. I know the former is not the case, I will assume the latter isn't either. If you seriously suspect an intoxication has occurred, get to the doctor. Let him decide what should be done.

Venomous Animals of the World

Chapter **1**

SOME VENOMOUS

MINOR
ANIMALS

The red sponge (*Microciona
prolipera*), capable of causing discomfort on contact. Relatively little is
known about the toxin produced by
these organisms.

he principal groups of venomous animals will be discussed in the chapters that follow. With the exception of the snakes, which require six chapters, each of the logical divisions, coelenterates, fish, amphibia, mammals, insects, lizards, arachnids, and molluscs, will be covered in individual chapters. In this first chapter we will touch briefly on some other groups of interest.

MICROSCOPIC ORGANISMS

Many microscopic animals, of course, produce and secrete or expel chemicals that are clearly toxins, substances that are dangerous for other animals to have inside of them. The toxin secreted by Botulinus B bacteria may be the most powerful toxin known to man. The toxin secreted by the anaerobic bacterium *Clostridium tetani* is what causes the damaged nerve tissue we call tetanus or lockjaw. But, microscopic organisms and their secretions, even their lethal secretions, are not what this book is about. We mention them here in passing only to create a context for our further discussion. Toxins start far down in the animal kingdom and below it among the bacteria. In their most primitive forms we call them disease.

SPONGES

Although not very much is known about the subject, some sponges (phylum Porifera) may impart a truly venomous substance. (Halstead has aptly labled our present level of ignorance on this subject a "vast, uncharted biochemical sea.")

The symptoms of sponge poisoning are essentially dermatological. They begin with a redness, which is followed by local swelling, skin eruptions, and blisters discharging a purulent fluid. The lesions may persist for weeks or even months. The red sponge (*Microciona prolifera*) from the northeastern United States is said to impart a feeling of stiffness in the finger joints (Halstead, 1965).

There is disagreement as to whether the symptoms are the result of chemical or mechanical irritation. (It is probably chemical, a primitive venom.) Structurally, many sponges (but *not* all) contain silica or calcium carbonate spicules imbedded in their substance. It is believed they are there for support. Presumably these spicules could cause mechanical irritation. They could also open tissue and allow venom to enter it. Sharp, penetrating mechanical devices associated with venoms are not unfamiliar to us—fangs, for example. When aqueous solutions of a number of sponge extracts were tried on experimental animals, the subjects exhibited a variety of reactions and many died.

There can be little doubt that sponges can secrete or discharge toxic substances that are irritating on contact. It can be assumed, but only that, that they are defensive in nature. The mechanics of the production of these substances, their chemistry, and their distribution on or in the body of the sponge are not known. Neither is it known precisely how they work on the organism that comes in contact with them. The fact that these substances will kill laboratory animals is more than slightly interesting. Here is a whole new area for investigation, for very little has been done in the study of poriferatoxins. (It is not beyond the realm of possibility that the sponges do not produce a toxic substance but rather adapt material they absorb in their primitive feeding system. Such patterns are not unknown.)

A typical case history involving sponges occurred among members of an underwater research group investigating Port Willunga Reef, St. Vincent Gulf, South Australia. One of their number was diving with an aqualung at a depth of 30 feet in the vicinity of the reef (the year was 1960), when he spotted a large, dark sponge chocolate-brown-bluish in color. It was about 18 inches across and almost the same in height. He cut the sponge loose and surfaced with it. Other members of the team manhandled it aboard the boat. Later that night the diver's hands were sensitive and painful. By the following day he could hardly hold a pen to write. On the second day he sought medical attention because his hands had become blotchy and swollen. It took five days for his hands to recover.

Others aboard the boat who handled or touched the sponge were also affected. One woman who touched it with her bare feet suffered extremely painful effects for several weeks and required extensive medical care. A nine-year-old boy suffered so much pain that he was crying piteously by the time the boat landed a few hours later. A photographer in the group handled the specimen freely and felt no immediate effects. The next day, however, the itching and pain started and spread to other parts of his body that had not come in contact with the sponge.

Everyone who handled the specimen seemed to suffer some ill effect, although all recovered. The sponge, they said, was slimy. We know little about what actually did happen to them in terms of chemistry and medicine. We should know more.

STARFISHES AND OTHER ECHINODERMS

The phylum Echinodermata is a large and interesting group of animals represented by the starfishes, the brittle stars, sea urchins, sea cucumbers, and related animals.

The class Asteroidea of the echinoderms contains the true starfish. It is known that some species, at least, produce a toxic slime in their epidermis that makes them both dangerous to eat and painful to touch. One species, the recently much-publicized crown-of-thorns starfish (*Acanthaster planci*), is believed to be truly venomous and to cause systemic reactions in man. An extremely painful wound can result from a spine breaking the tissue. There may be swelling, discoloration, a local rash, numbness and, reportedly, persistent vomiting and some degree of paralysis.

Again, we know very little about the mechanisms of this form of contact envenomation. It is assumed that its purpose is defensive. Cats have been known to die after eating certain species, and it may be that all starfish are to some degree or other repellent to would-be predators. They do not create any kind of a public health problem, and extended research into the makeup of their venomous properties may be slow in coming. As we shall be seeing there are many areas of venomology where an urgency can be demonstrated. Research tends to concentrate in those areas when research dollars are in short supply. This can be a short-sighted attitude, however. The increase in our understanding of any toxin is an increase in the understanding of all, even if the truth of that statement is slow in making itself apparent.

There is some talk about brittle stars (class Ophiuroidea) being poisonous or venomous, but we are dealing here with an animal remote from the experience of man and with no record of threat or intimidation. Nothing is known about the subject really, and we can only make the passing suggestion that something may someday come to light.

Sea urchins (class Echinoidea) are another story. They are venomous in two different ways—through specialized spines, and through small seizing organs scattered over the surface of the urchin known as *pedicellariae.*

Most sea urchin spines are not venomous. Some, however, are extremely sharp and brittle and can break off in the flesh of a victim, causing great pain and distress. Certain spines on certain species are apparently equipped with venom glands that encase their tip and rupture on contact. Halstead names three species as examples: *Phormosoma bursarium* (no common name provided) is a widespread urchin found from Hawaii west to Africa. It has short spines and those near its mouth (on the underside, as with all sea urchins) have their tips sheathed in a venom sac. *Araeosoma thetidis*, the tam-o'-shanter urchin of New Zealand and Australia, also has venomous spines, as does *Asthenosoma varium*, the leather urchin of Indonesia, the Indian Ocean, and the Gulf of Suez. In each case only a limited number of spines are known to be venomous.

The pedicellariae are small devices found scattered between the spines, and some forms are venomous. They normally have three jaws, each of which may have its own venom gland. While the spines are almost certainly defensive, the small mouth-like devices can be used for grasping and envenomating prey. It is highly unusual for an animal to have two totally different kinds of venom apparatus. Usually, even when there are two distinct functions, one device will serve both.

The crown-of-thorns starfish (*Acanthaster planci*). This species is known to be venomous. The venom is thought to be produced in cells within the membrane that covers the spines.

Where the spines of venomous sea urchins penetrate and cause a true intoxicatoin there is immediate pain. Swelling and discoloration follow the first burst of burning pain and then a deeper aching sets in. Systemic effects recorded are: slight motor paralysis, swelling of the face, and pulse irregularities. The toxin is not known to have been fatal to anyone, but deep, secondary infections have resulted.

All long-spined sea urchins should be avoided. Even fairly heavy canvas gloves and shoes provide inadequate protection. Short-spined tropical species may be able to inflict a wound with their jaw-like pedicellariae, and this should be kept in mind. Urchins are for looking at, not touching.

While not very much is known about the venom of sea urchins, enough has been reported to indicate caution. The injuries reported could have been more mechanical than chemical, but then again they may not have been. There is no doubt that the spines of some sea urchins may be among the longest envenomating devices known.

MARINE WORMS

The various marine worms may be toxic to some degree, although not very much is known about this aspect of their morphology or behavior. The phylum Platyhelminthes, the flatworms, contains some species that are probably poisonous to eat but not venomous within the guideline we have set down for ourselves. It is conceivable that the slime secreted by some, probably as a repugnatorial device, could be harmful, but that is not known.

The phylum Rhynchocoela, the ribbon worms, may contain species with a sharp stylet that bears a true venom. Presumably this stylet, set on the end of an extensible proboscis, can be jabbed into anyone attempting to handle the animal. It could be used to subdue prey, or it may be entirely defensive—very little is known.

The phylum Annelida contains the segmented worms. Some of the marine annelids are apparently equipped with a venom. In some species there are setae or bristles, as they are sometimes called, and these reputedly can impart a rash and generally localized inflammation. Other species have small, fang-like jaws on the end of a retractable proboscis. At least some species have venom glands associated with these jaw/fangs, and a painful nip is possible if the animal is carelessly handled. In both cases, contact with setae or bites by venomous jaws, anything beyond the pain, itching, and annoyance of a highly localized nature, would be unlikely.

These, then, are a few animals we will *not* be discussing in the chapters that follow. They are given here as examples of the

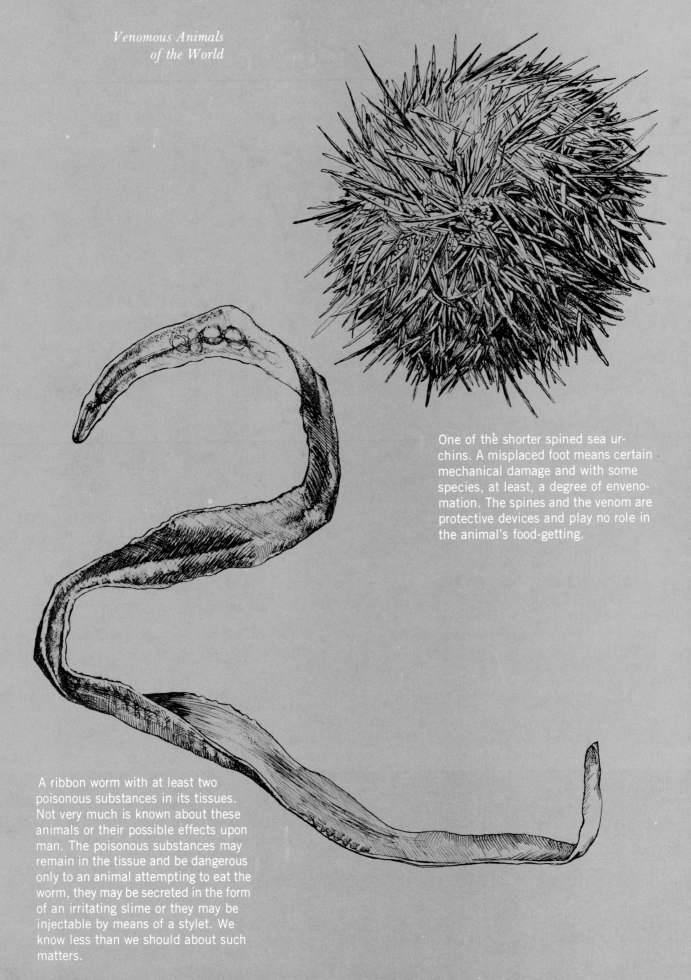

One of the shorter spined sea urchins. A misplaced foot means certain
mechanical damage and with some
species, at least, a degree of envenomation. The spines and the venom are
protective devices and play no role in
the animal's food-getting.

A ribbon worm with at least two
poisonous substances in its tissues.
Not very much is known about these
animals or their possible effects upon
man. The poisonous substances may
remain in the tissue and be dangerous
only to an animal attempting to eat the
worm, they may be secreted in the form
of an irritating slime or they may be
injectable by means of a stylet. We
know less than we should about such
matters.

minor venomous forms that inhabit that part of the animal world with which we have limited contact. Many more, of course, are poisonous to eat, but that is a subject for a different book.

Do any of the forms briefly mentioned in this chapter represent a health problem? Aside from an occasional accident with a sea urchin, not really. Such accidents are of very little consequence when compared with the real subject matter of this book, as we shall be seeing.

No one knows (it is doubtful that anyone will ever know) just how many animals have venom glands. Very often closely related animals will show marked differences in this regard. We should not overlook the fact, however, that our reaction to or interaction with a venom is only one measure by which it can be evaluated or even identified. A small bristle worm crawling across a coral reef in some remote sea in search of prey may be as deadly to the small animal whose substance it seeks as a king cobra would be to us. Our relationship to a venom, the degree to which we react to it, is only one very small yardstick. There are many others.

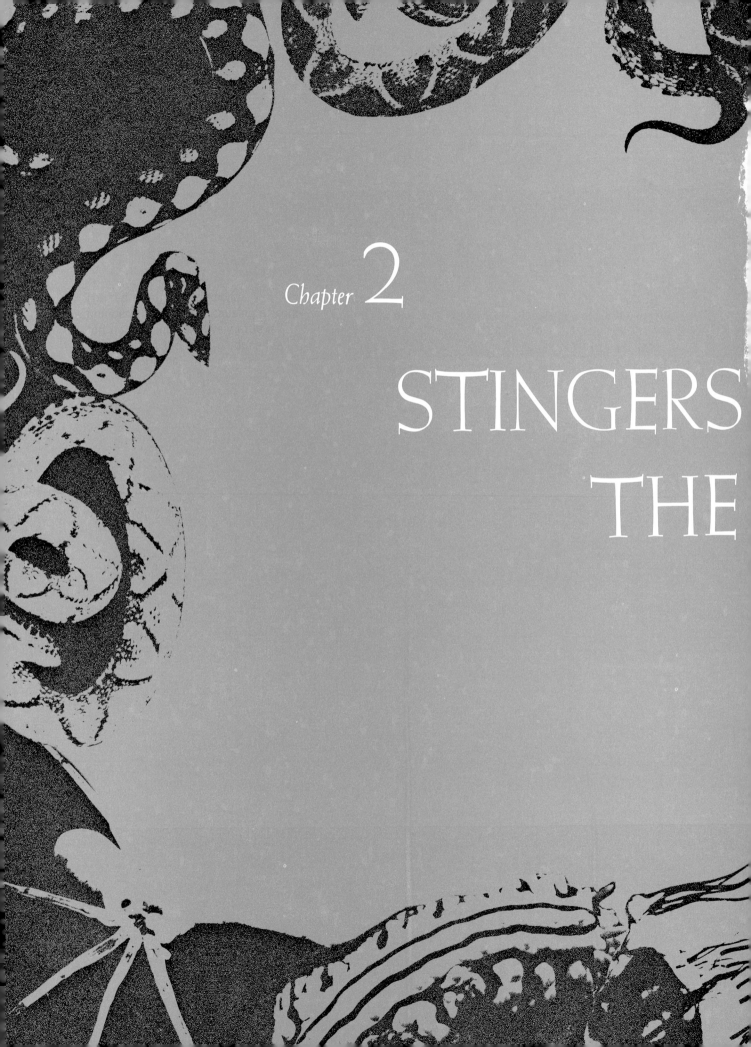

Chapter 2

STINGERS
THE

IN THE SEA – COELENTERATES

Fire coral *Millepora*, a source of
discomfort to many swimmers but not
venomous enough to be truly dangerous. Localized effects can be variously
severe.

he coelenterates or cnidarians are simple animals. They are the lowest evolutionary group of animals to have true tissues and a fixed shape. They have one central, internal cavity, or *coelenteron*, and this leads to a simple mouth. What is of interest to us is that their epithelial layers are equipped with hundreds of thousands of highly specialized cells called *nematocysts*. Some are annoying to man, some lethal. (Nematocysts are commonly referred to as cells although they are not true cells at all. They are technically *cell organoids*.)

The coelenterates, in a great variety of shapes and sizes, and differing widely in habit, are divided into three classes:

The Hydrozoa are superficially plant-like in appearance. They often look like small plumes growing in shallow water. They generally have a medusa, or free-floating stage, although there are exceptions. An important exception occurs in the hydrozoan order Siphonophora—this order contains a planktonic colonial medusae popularly called the bluebottle or Portuguese man-of-war. The Portuguese man-of-war is often thought of as a jellyfish. It is not a jellyfish, though, since it is not a single animal but a group of animals living in a highly efficient and highly mobile colony.

The Scyphozoa contain the medusae or true jellyfish. The 200-odd species are marine and generally pelagic. It is a class we will discuss at some length in this chapter.

The Anthozoa is the third class of coelenterates, the sea anemones, corals, and alcyonarians. Animals in this class have no medusa stage, and it is difficult to think of them as being in the same group with the fragile jellyfish—unless one examines their essential structures. A coral reef may not resemble a jellyfish, but the animals that build it do. The alcyonarians are the least known of the Anthozoa. They are the soft corals, thick masses of fleshy material imbedded with particles of limestone.

All of the coelenterates have tentacles equipped with specialized organoids that qualify them as truly and seriously venomous. As far as is known their venoms are proteins. Unlike most of the other animals we will discuss in this book, the coelenterates do not have their venom in a single repository—or even in several. The startling and occasionally lethal power of these animals is distributed by hundreds of thousand of microscopic capsules each with its own injection device. In one jellyfish—*Chironex*—there may be 80,000 nematocysts to the square centimeter—516,000 per square inch.

HYDROIDS

The class Hydrozoa contains a variety of stinging animals most of which are of little consequence to man. Hydroids, usually more common in temperate and cold zones, are often colonial. Two types of polyps or zooids combine to form the creature as we see it. The nutritive polyps feed the colony and the gonangia, or reproductive polyps, assure future generations. It is the nutritive member of the colony that makes the organism venomous, for nematocyst-equipped tentacles are used in capturing food.

Although the true coral belong to the third class of coelenterates—the Anthozoa—the Hydrozoa contain the so-called hydroid corals. They are prominent features on large reefs building up on top of the true corals that form the foundation on the seabed. They are frequently blade-like and branching. Those that concern us here are called fire or stinging corals. The fire coral *Millepora alciornis* is a common animal in the waters around Florida and is widely distributed throughout the West Indies and the Caribbean. The nematocysts of this species are capable of penetrating human skin, resulting in a whole series of reactions. There can be intense local pain, papular eruptions, and pustular lesions. Healing can be slow and not uneventful. The species is now being studied, and it is not clear whether all of the toxic material comes from the nematocysts or whether some comes from surrounding tissues as well.

A congener from the other side of the world is *M. dichotoma*, found in the Red Sea and the western Pacific Ocean. Reactions in man to contact with the two species are similar. The toxin extracted from the Atlantic species appears to be more stable and in laboratory mice creates immunological reactions. Much work remains to be done on the fire coral.

A hydroid found in Hawaiian waters—*Syncoryne mirabilis*—can cause severe dermatitis. In some victims the lesions have become infected and healing has been slow and uncomfortable. An urticarial reaction, however, is not the full story. Another Pacific species is suspected of causing a much more severe syndrome that includes abdominal cramps, chills, high fever, general "sick" feelings, fear and other psychological disturbances, and diarrhea. As we examine the potential of some of the other coelenterates such a syndrome will not seem surprising.

THE NEMATOCYST

Most of the coelenterates have their venomous potential limited to their stinging cells. Before going on to discuss the animals themselves, it would be well to review these minute mechanisms. The nematocyst, or nettle cell, comes in a number of forms (perhaps as many as 20), but they are essentially the same as far as we know. The organoid is a sealed unit with a trigger or sensor called the cnidocil. The nematocysts can be set off mechanically by contact with these triggers, although chemoreception can play a role as well. Personally, I do not think there has been a reliable quantification of these factors. There appears to be evidence that discharge can be under a form of central control, perhaps hormonal-like. This shows a rather advanced level of organization.

When the animal is near its victim the nematocysts discharge.* A thread tube coiled within the venom capsule shoots out with astounding speed and power. It is variously reported that an operculum flies open as the thread tube emerges and that the tube penetrates its own capsule wall. In experiments tubes have been ejected with enough force to penetrate heavy laboratory gloves. There is no indication of nerves running to these nettle cells. The fact that they have their own receptors, both chemical and mechanical, and that they can discharge themselves makes them possibly the lowest form of sensory-motor system yet recorded. Whether pressure built up within the nematocyst capsule gives the tube its force, or whether there is an elastic effect, is not clear. How a central control would work is even more puzzling. The chemotaxis aspect of nematocyst discharge, though, seems indisputable. Flesh triggers the cells much more readily than glass or plastic.

>*Tentacles floating free in the ocean can sting, as can those washed up on the beach. Even when the bulk of the parent animal has been destroyed, the nematocysts work.

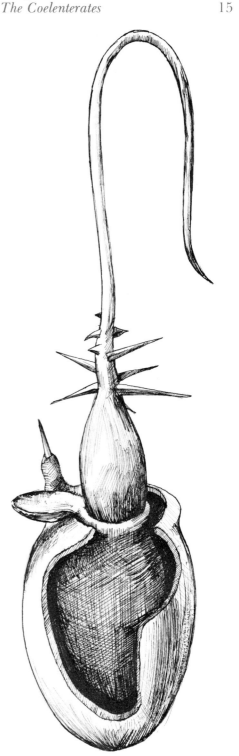

The discharged nematocyst. The thread-like filament bearing the noxious chemicals has burst through the lid or operculum. The small single barb at the left of the operculum is the trigger that when mechanically disturbed discharges the stinging unit.

Of special interest is the use of nematocysts by other animals. Some reports have the octopus *Tremoctopus violaceus* picking up Portuguese man-of-war tentacles and holding them in their own tentacles for use against other animals. Early in 1968 bathers off New South Wales, Australia, reported numerous incidents of unpleasant stingings. The culprit was an apparently harmless blue sea slug that came in with the tide. It was determined rather quickly that small slugs (nudibranch molluscs) of the genera *Glaucus* and *Glaucilla* had been feeding on coelenterates and were capable of storing undischarged, undigested nematocysts and employing them in their own defense. Other nudibranchs are known to have this strange capacity. Long before the first man-ape picked up a stick or rock, small, shell-less mollusks in the sea had learned to pick up venomous capsules and make them their own defensive mechanism. In the strange world of the sea this is one of the strangest stories yet told.

The nudibranchs, marine molluscs without external shells, produce a toxic slime that covers their entire bodies. It is paralytic and deadly to some organisms that might be tempted to feed on these slow-moving animals but the nudibranch is not equipped to externalize the toxin or force it on another organism that keeps its distance. The vivid markings of these molluscs are protective—they warn other animals away so that mistakes are not made.

Whatever kind of nematocysts a coelenterate might have, the method of intoxication remains the same. The tubes thrown out from the venom capsules stick into the victim and venom is discharged through them. The result can be anything from a very mild rash or burning sensation to death in a few minutes. A number of factors, of course, determine how serious the injury will be. The type of nematocyst and the type of venom are important, but ultimately the identity of the animal is the most significant factor. The amount of bare skin exposed and therefore the number of discharged tubes encountered can be of importance. Then there is the matter of individual sensitivity. Some people are more sensitive to foreign proteins than others, and an episode of envenomation may be affected by this factor.

PHYSALIA, OR PORTUGUESE MAN-OF-WAR

For a long time fatal encounters with "jellyfish" were attributed to *Physalia*, the Portuguese man-of-war. This animal was almost universally thought of as deadly, and little attention was paid to true jellyfish, beyond, of course, avoiding contact with them.

The Portuguese man-of-war is not a jellyfish, as we have seen, but a Siphonophor—a medusa-like hydrozoan that is atypical of the class as a whole. It is capable of causing severe discomfort, but its venom is rarely if ever deadly. Encounters, though, can be distressing since a specimen of *Physalia* with a float 12 inches long may be trailing 75-foot tentacles in the water below. There is at least one case of an elderly man swimming off Florida's east coast and encountering a *Physalia*. He was badly stung and suffered excruciating pain. It was severe enough to cause a heart attack and, although he was originally thought to have been killed by the animal's venom, it was later shown that heart failure unrelated to envenomation, except insofar as it was pain-induced, caused his death. Another Florida case involving the death of an eight-year-old girl has been corrected in the records. *Physalia* had been suggested as the cause of death. Such was not the case.

The venom of *Physalia* is extremely complex. It is likely that its toxic activity is the result of various components acting together rather than independently. Among the components identified so far are phospholipases A and B, several neutral lipids, various enzymes, and peptides. Altogether it is a remarkable substance. Scientists working at the University of Miami have determined that *Physalia* toxin blocks nerve conduction, giving a severe systemic syndrome up to and including shock and collapse.

The question occurs again and again as to whether *Physalia* intoxication can lead to human death. The confusion is compounded by the appearance of early medical literature stating categorically that it does. Many case histories were recorded. Later it was shown that the victims had not been stung by *Physalia*, but by certain cubomedusae jellyfish of the class *Scyphozoa*. Nonetheless, the legend was established and it is very difficult to correct it. It should be enough to suggest that for certain people under certain conditions, a serious encounter with *Physalia* could possibly lead to death. That appears unlikely, however, particularly since at certain times of the year hundreds of *Physalia* victims are reported along Australian beaches. People form a line waiting for medical treatment. All apparently survive.

SEA WASPS AND OTHER JELLYFISH

If there is a question as to the effects of *Physalia* venom, there is none at all about those of jellyfish of the family Chirodropidae. Within this family there are at least two of the most seriously venomous creatures on earth—the incredible sea wasps or box jellies.

The box jellies *Chironex fleckeri* and *Chiropsalmus quadrigatus* are decidedly deadly to man, the former even more so than the latter. Up until recently they were little known and their extraordinary potential was not understood. The Portuguese man-of-war, as we have said, was erroneously blamed for the deaths now known to be attributable to these remarkable animals. Few animals on land or in the sea have such devastating potentials.* In 1965, Cleland and Southcott published data on 53 fatal jellyfish encounters that had occurred between 1884 and 1958. The average age of the victim, where ages were recorded, was fourteen years. In 37 cases death occurred very quickly. In 28 cases death came in less than ten minutes.

Chironex fleckeri, certainly the most dangerous jellyfish in the world and one of the most dangerous animals, can be up to 30 feet long with its tentacles extended. It has toxin that may be six times as potent as its lethal runner-up, *Chiropsalmus quadrigatus*. So far only one species has been assigned to the genus *Chironex* and that is this animal from northern Australia waters—notably Northern Territory and North Queensland. There are at least three members of the genus *Chiropsalmus* that are known to be dangerous—*C. buitendijki*, a species ranging from the Malay region down to Australia; *C. quadrigatus*, a species of particular concern, found ranging from Northern Australia to the Philippines and west into the Indian Ocean; and *C. quadrumanus*, a very widely distributed species found near Australia, in the Indian Ocean, and in the western Atlantic at least from North Carolina to Brazil, and undoubtedly in areas not yet reported.

*During World War II in the southwest Pacific there were reports of men on life rafts suddenly dying and fish were often blamed. Even then Dr. Sherman Minton suspected a form of jellyfish not yet identified.

The box jelly *Chiropsalmus quadrigatus,* quite possibly the second most seriously venomous animal in the sea. It's toxicity is surpassed, in all probability, only by that of the sea wasp *Chironex fleckeri.*

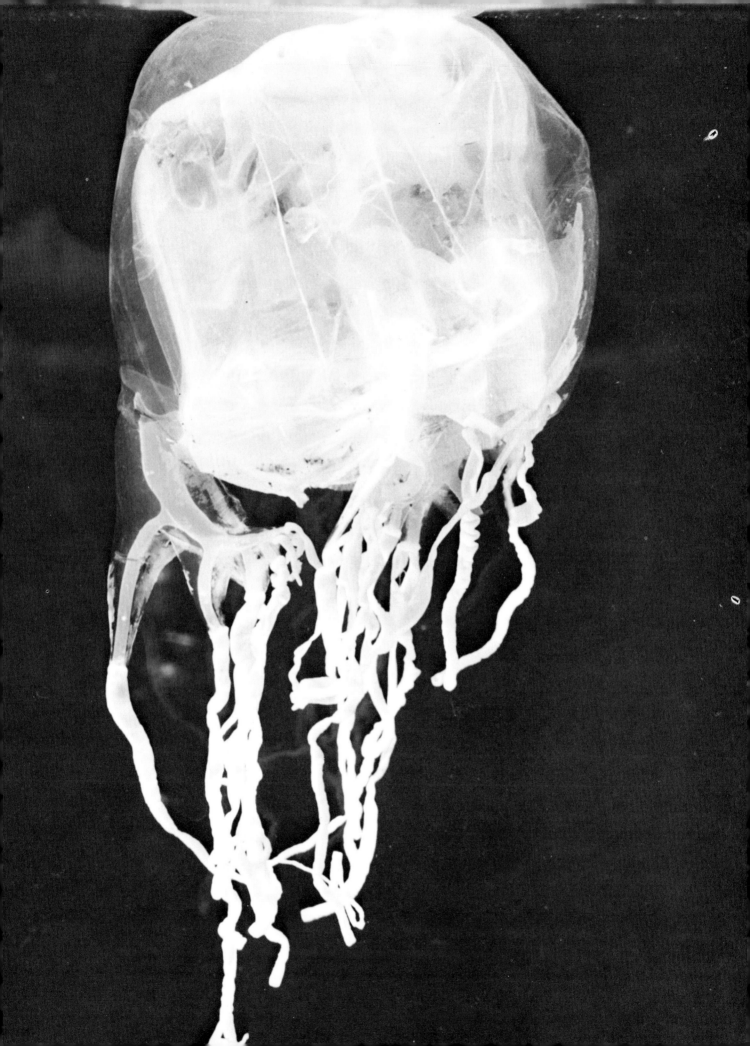

The two sea wasps or box jellies *Chironex fleckeri* and *Chiropsalmus quadrigatus* (both are usually called sea wasps, the latter alone properly box jelly) are apparently responsible for many heretofore unexplained deaths in Australian waters —including not only those blamed on *Physalia*, but those without any previous explanation at all. This fact applies to other areas in that part of the world as well. For example, in the extensive data assembled by Halstead, four people were stung by "jellyfish" in the Philippines in 1927 and two died; two people were killed by "jellyfish" near Darwin, Australia, in 1955; ten out of eleven *Physalia* victims in Australian waters between 1937 and 1951 were said to have died. A fatal *Physalia* victim was listed for New Guinea in 1943-44, and "jellyfish" fatalities were reported for North Queensland in 1949, and two for Palawan Island in the Philippines in 1951-53. There are many more recorded. Some sources suggest 50 fatalities for Australia alone in this century. All of this, apparently, our two sea wasps have done. This and much, much more.

The deadliest of all jellyfish, the sea wasp *Chironex fleckeri*. At least thirty-nine people have been killed in Australian waters, and many more around the Philippine Islands.

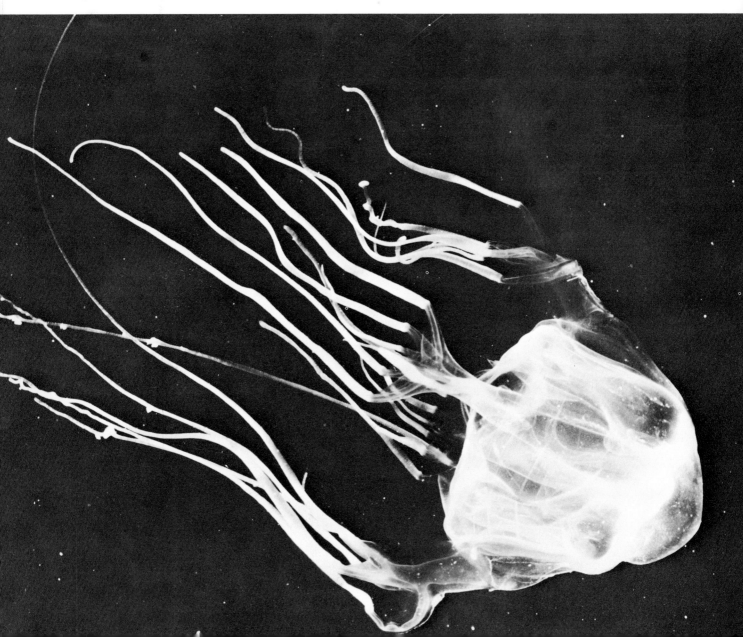

The two sea wasps we are discussing are superficially similar animals, although nematocyst density is apparently much greater in *Chironex* and its toxin is also believed to be more potent (by six times as we have suggested). The differences between the two sea wasps' toxins are being carefully studied. In laboratory experiments, researchers have found that both types trigger acute hypertension by causing vasoconstriction. Cardiac irregularities follow, and prior to death there are severe oscillations in arterial pressure. Even when diluted 10,000 times *Chironex* venom can kill mice by asphyxiation. Well-defined lethal fractions of both venoms have been isolated and used in extensive toxicity experiments. The results being gathered are not academic. The two jellyfish are so extraordinarily dangerous to man that the information will undoubtedly help save human lives.

The venom of these most lethal of jellyfish is obtained in two different ways. The animals can be "milked" by getting them to discharge their nematocysts into an amniotic (fetal) membrane. The venom is collected on the far side where the tubes have emerged and discharged their dangerous loads. Alternately, entire tentacles have been ground up and the venom fractions separated by mechanical and chemical techniques. Two lethal fractions have been isolated in *Chironex* venom and one has a much greater molecular weight than the other.

When seas are heavy and the weather uncertain cubomedusae gather offshore and offer little threat to man. (This is not true for commercial fishermen hauling nets, however. Reportedly, these animals discharge more readily and more powerfully when touched by anything smelling of fish oil. This would make them a particular hazard to fishermen.) When the weather is calm, cubomedusae can gather in shallow water, often along well-used recreational beaches. It is this characteristic that has placed many reports in the medical literature—like that of a five-year-old boy who died near Cardwell, North Queensland, on a calm, sunlit morning in January, 1955. He was paddling no more than *six feet* from shore when he screamed, ran from the water, and died on the warm sands where his family was sitting. The jellyfish was of a previously unnamed species. It was given the name of *fleckeri* in honor of Dr. Hugo Flecker, the Australian physician who had spent 20 years researching the cause of unexplained human fatalities in Australian waters.

Before the identity of the sea wasp was determined, Dr. Flecker had argued long and hard against the assumption of a culprit called *Physalia*. I have a typescript he used for a speech given somewhere in Australia prior to 1955. I do not have the details of when or where it was given, just his typed notes. In them he says:

For years past, similar fatalities have occurred off the north Queensland coast, where death has been particularly prompt, usually in young healthy people, almost always between twenty and forty minutes. Although in no case has the unmistakeable small bluish bag of the *Physalia* been seen, invariably has it been announced by the lay press particularly that the *Physalia* is the cause. The jellyfish has even been seen and noted to be a large colorless mass, several inches, perhaps, in diameter. Post mortem reports have concluded from the large lash-like weals that the *Physalia* is responsible and do not go further to seek the culprit.*

Of the 30 species of known stinging coelenterates in Australian waters fewer than a dozen are actually troublesome to man. Only three create problems of any importance. The first two we have discussed; they are the sea wasps. But, there is a third and although its effects are nonlethal it did give cause for concern on the public health level. Dr. Flecker sought the answer to this riddle as well. In 1952 he proposed the name "Irukandji stinging" as a shortcut to the many names then being given the syndrome in the popular and technical press. Irukandji is the name of an aboriginal tribe whose hunting territory included areas where the stingings were particularly common. The name was adopted and indeed did facilitate the filing and recovery of data.

The early reports of this particular stinging syndrome date from the 1920s and 1930s but it was not until 1961, four years after Dr. Flecker's death, that the animal was identified. It is the *Carubia barnesi***, a small animal no more than an inch and a half in length. The symptoms, though, would seem to come from contact with a much larger animal. There is a sharp sting, rather like that of a wasp. A prickling, tingling effect can persist for one to three hours. Sometimes after 20 minutes have elapsed systemic symptoms appear. The victim may be overwhelmed by shock and collapse. There can be severe headache, intense cramps in the back muscles, the abdomen, and the limbs. There may be retching and vomiting, and the agony may continue for hours. There is profuse sweating, labored breathing, disturbed vision, and general depression. The syndrome is self-limiting, however, and recovery is usually complete in a day or two.

*Dr. Flecker's *Physalia*, or blue-bottle, is very similar to our Portuguese man-of-war. Over the years Dr. Flecker apparently became frustrated by the resistance of his peers to his views. As it turned out he was almost 100 percent right.

**Named after Dr. J. H. Barnes who finally identified the culprit by experimenting on himself and members of his family.

An antivenene for the sea wasp has been developed in Australia. However, the venom itself peaks so quickly there is seldom time to locate and utilize a chemical treatment. Greater hope is to be placed in a program of immunization now being developed.

There are other jellyfish capable of seriously affecting an organism even as large as man. The sea nettle *Chrysaora quinquecirrha* is found around the world—from Massachusetts to Japan. Although nonlethal its venom has a number of things in common with that of *Chironex*. They are both large proteins with cardiotoxic, neurotoxic, hemolytic, and dermonecrotic properties. Interestingly enough, scientists working with this jellyfish at the University of Maryland School of Medicine found that its venom had properties present in the poison of some salamanders and fish. Both substances work similarly to block signals and interfere with the action-potential in excitable cells. We have much to learn.

The sea nettle *Chrysaora quinquecirrha*, a true jellyfish equipped with hundreds of thousands of stinging cells. Another species with a venom powerful enough to cause skin lesions but not toxic enough to cause severe or prolonged systemic reactions.

Cyanea capillata is another jellyfish worth noting. The largest known jellyfish, it can be almost ten feet across the bell with tentacles trailing 120 feet below. This giant sea blubber or lion's mane has as enormous range in both the Atlantic and Pacific Oceans. Stings are not fatal but can cause great discomfort; whealing, difficulty in respiration, frothing at the mouth, and loss of consciousness. Clearly, an encounter with a mass of tentacles 120 feet long could be disquieting.

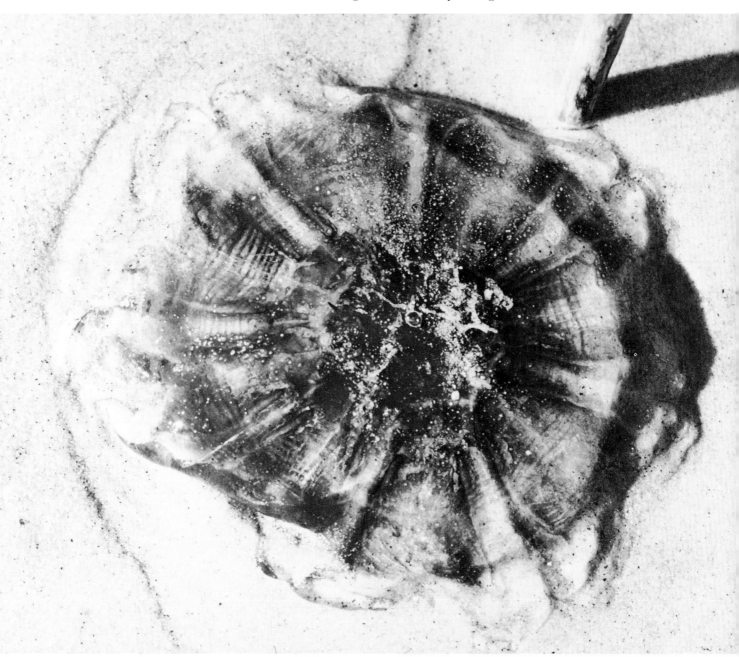

The jellyfish *Cyanea capillata,* a common, widespread, and mildly venomous species. Its stinging mechanisms closely resemble those of deadly species but it is not believed capable of killing organisms as large as man.

In discussing animals as obviously dangerous to human beings as sea wasps and as unpleasant to encounter as *Physalia*, there is an understandable tendency to think of them only in terms of their relationship to man. That, of course, does not lead to an understanding of the animal, nor is it really the purpose of this book. The stinging cell organoids of the coelenterates are what enables them to fill their niche and survive as hunting animals.

The fossil record of the coelenterates is very poor, except for the corals and fire corals. Jellyfish do not leave much of an impression on stone. We do not have an actual record, therefore, of the history of this group of animals. We are faced with a fait accompli—about 20 different kinds of nematocysts, the functions of only a few of which are really understood. Still, a family tree for the jellyfish and related organisms has been postulated based on existing structural and chemical relationships. They are very ancient animals.

As we will ask ourselves again and again in this book, "Why do certain animals develop powers so extraordinary, so out of proportion to what would seem to be their needs?" The sea wasps are capable of destroying a 200-pound organism at its peak of health and activity, in a matter of a very few minutes. The jellyfish feeds on small sea animals that could be stunned into unconsciousness, or killed, with only a small fraction of this potency. As for defense, there are animals that eat jellyfish and their relatives; not only are these predators not driven off by a battery of discharging nematocysts, they are capable of absorbing and utilizing those that have not been fired.

I have never believed that nature wastes itself on useless structures and behavior. If a system or organ is devised that fails to fill a survival need, it is either reabsorbed into time and evolution, or the host animal vanishes. If this is true, why the sea wasp? Why such truly extraordinary powers? We will draw many such analogies, but the kind of power packed into the five pounds of flesh that is a mature *Chironex* jellyfish, when compared with the kind of prey it seeks, is analogous to a human being hunting a squirrel with an atomic cannon. It is a supreme case of overkill, or at least so it appears.*

*Minton suggests we know so little about this animal's niche in the highly competitive marine environment that we can draw no conclusions. He writes, "It may be that 80,000 nematocysts per centimeter of tentacle is just what *Chironex* needs to exploit a particular food source. It is probably pure accident that it can kill a man in a matter of minutes."

Chapter 3

THE

MOLLUSCS

The Hebrew cone (*Conus ebraeus*), one of the four to five hundred cones, all venomous, some deadly. These specimens were collected on Australia's Great Barrier Reef. This is believed to be a sublethal species.

he molluscs, phylum Mollusca, are a widely divergent and particularly interesting group of marine, aquatic, and terrestrial invertebrates. The phylum contains the seashells, often coveted collectors items, that are widely known and deeply appreciated for their beauty of form and color. It is surprising how few people seem to know that a number of molluscs are deadly to man.

The structure of the phylum is subject to some considerable debate. The number of mollusc species is variously stated to be anywhere from 45,000 to 80,000, depending on the authority. The number of classes into which the phylum should be broken down is also subject to discussion. Lord Rothschild gives it as seven and some other writers agree. Halstead suggests that five is the usual number. I prefer to use six. Only two classes (two upon which all writers seem to agree) contain animals appropriate to the subject of this book.

The six classes of Mollusca, if you will, are:

Amphineura—These are the chitons. Some species may be poisonous to eat but none are believed to be venomous. Very little seems to be known about even their poisonous qualities.

Scaphopoda—These are the tooth or tusk shells. None are venomous, certainly, and none are thought to be poisonous.

Gastropoda—From our standpoint this is the class of maximum interest, and it is easily the largest class of Mollusca. There are probably more than 33,000 species, although this is admittedly an estimate. The gastropods are univalved, having either a single shell or no external shell at all. They are found on land and in both fresh and salt water. All of the snails and slugs belong to this class and the animals themselves are generally asymmetrical and formed into a coil or spiral. They usually have a distinct head with tentacles and eyes. Movement is achieved by a large, fleshy foot. A number of species are seriously venomous.

Pelecypoda—These are the bivalves, the molluscs with two shells. Included are the scallops, mussels, oysters, and clams. There are about 11,000 living species and none are venomous, although a good many are seriously poisonous when eaten. They are of enormous importance to man as a food source and are harvested by the millions of tons each year.

Cephalopoda—The "head-foots," the cuttlefish, nautilus, squid, and octopus are included in this class. There are about 650 species still living and most have no external shell. This class is the only one besides the Gastropoda that contains venomous molluscs. One and possibly a few are deadly to man.

Monoplacophora—This is a relatively small class of deep-sea molluscs that are "primitively segmented" in their body organization. Little is know about them but none are suspected of being venomous.

The molluscs generally are unsegmented animals. They are soft-bodied and the outer surface of the mantle, the soft tissue that covers much of their body, usually secretes a calcareous shell. It is quite usual for uninformed people to think of the molluscs as just seashells without any regard for the animals that live inside. Indeed, some people have apparently thought of seashells as mineral deposits of some kind rather than as hard structures created by individual animals for their own protection. Halstead has identified the molluscs as ". . . the largest single group of biotoxic marine invertebrates of direct importance to man." This is primarily true because of the millions of tons of shellfish harvested annually by man and the numbers of people poisoned each year through eating these animals. That aspect does not concern us here, however, as no venom is involved. The venomous molluscs offer a study and a concern more than sufficient to our needs and interest in this book.

Within the class Gastropoda, the cone shells belong to the family Conidae in the suborder Toxoglossa. With good and sufficient reason, Toxoglossa means "poison-tongue." There are somewhere between 400 and 500 different cone shells and as far as is known all are venomous, although the degree to which they are so may vary considerably.

The cone shells are extremely popular with collectors. They are often very beautiful and one species, at least, *Conus gloriamaris*, the Glory-of-the-Seas Cone, is very valuable. Specimens have sold for as much as $2,000. Advanced shell collectors prefer to collect live specimens, and this has brought a great many people into contact with the living animal itself, not always with the happiest of consequences. The cones, which are usually associated with coral reefs—generally located in shallow water in tropical seas—are often accessible in intertidal zones. Although many species hide under rocks or bury themselves in the sand, it is still common to find clusters of them at low tide, often with a number of species intermingled. The ease and frequency of contact between man and animal has resulted in a considerable literature about these colorful and sometimes dangerous creatures.

The cones are divided into three groups, depending on food preferences. First there are the vermivorous species —the worm-eaters. A variety of marine worms are used as prey but mainly the polychaetes. Nowhere near as common or widespread as the vermivorous cones are the piscivorous species —fish-eaters. Both fish and marine worms are active animals, and the snail-like animal that is going to utilize them for prey must be equipped to stun or kill them almost instantaneously. The third group of cones consists of gastropod-eaters, which prey on other and perhaps equally venomous cones. In those cases where a venomous cone attacks a venomous cone the envenomating apparatus of the prey animal is apparently useless for defense. In fact, the venom of the cone shell is primarily a food-getting device and only of secondary importance for defensive action. The type of venom a cone has at its disposal is believed to be directly related to the type of prey preferred.

The fact that some conus venom contains digestive enzymes has strengthened the belief that the whole envenomating system evolved from a digestive gland. This does seem more than probable. But, as we shall see, the system is by no means restricted to use against organisms that a cone shell can digest.

In 1705 a slave woman on Banda Island (Indonesia) was hauling nets along the beach. Somewhere in the mass of squirming fish was a beautiful little cone shell we call the textile cone (*Conus* *textile*).* Whether the woman actually grabbed the shell or just brushed it is not clear, but in short order she was dead. The episode was witnessed by the Dutch naturalist G. E. Rumphius, who gave the literature of Europe its first eye-witness account of cone shell envenomation.

Ninety years later the British warship "H.M.S. Samarang" was cruising in the Indonesian archipelago when a sailor somehow came into possession of a living cone of an undetermined species. He was made seriously ill and the British naval surgeon in attendance, Arthur Adams, reported the incident. In 1848, another British sailor was seriously envenomated in the Moluccas by a court cone (*C. aulicus*). Again an official report was made. By that time the fact that the tropical Pacific and Indian oceans harbored venomous seashells was widely known. Yet, despite this somewhat startling knowledge, and the tremendous interest taken in cone shells generally because of their beauty and "collectability," little research was attempted. All of that has changed, however, and a wealth of literature attesting to the remarkable nature of these animals is now available.

The cones are largely nocturnal and active, at least near the shore, only at high tide. They glide along the bottom, across the sand, over rubble and debris, seeking prey. Although their forward movement is inexorable it is not swift. They compensate for their lack of swift grace with a highly unusual food-getting system—they harpoon their prey with toxic barbs.

The size of a cone shell varies greatly from species to species. A full-grown example of the pygmy cone, *C. pygmaeus*, may measure no more than 25 millimeters, while a specimen of the leopard cone (*C. leopardus*) may weigh several pounds and measure eight times the length of the pygmy. Interestingly, the size of the barb or harpoon with which a cone shell is equipped does not necessarily correlate with the size of the mature animal. Some of the smaller cones have the longest harpoons.

The entire venom apparatus of the cone shell lies in a compact mass in and around the animal's esophageal region. Furthest from the shell's aperture is a bulb with thick muscular walls. Some small amount of venom *may* be produced in the bulbs of some species although that is doubtful. The bulb is a hydraulic device, a means of propelling venom forward into a harpoon. The harpoon, barb, or radula tooth, as it is sometimes known, is "charged" prior to use and carries its own venom supply with it once it has been plunged into a victim. It is apparently the job of the bulb to force the venom forward into the tooth. It is not clear whether this is done in advance or just prior to actual use.

*At least one investigator, Endean, believes reports of *C. textile* being lethal are cases of mistaken identity. *C. geographus* would be more likely.

A fine tube runs forward from the bulb to a short section of the anterior gut known as the pharynx. This tube is the venom duct, a twisting, undulating device that, stretched flat, could be five times the length of the animal itself. It coils and twists around the other parts of the venom apparatus. It is within this duct that the bulk, perhaps all, of the venom is produced. One of the many mysteries surrounding the cone shells arises in this tube. The inner cells of the tube apparently produce minute venom bodies, and each of these in turn is apparently a venom factory. At the forward end of the duct, near where it joins the pharynx, the venom bodies are exceedingly small—in one species (*C. magus*) these bodies have been measured at 200 millionths of an inch in length. As the bodies migrate back through the tube, which they do as a matter of course, they grow in size. Back near the pneumatic bulb end of the tube they may be four times the size they were at the beginning of their journey. They also produce much more powerful venom as they grow and move. The venom from the rear end of the duct may be 20 times as potent as that at the front end. For a long time researchers puzzled over this seemingly contradictory fact. Why should venom be much more powerful away from the harpoon than near it?

Conus textile, the textile cone, one of the most seriously venomous of all the cone shells. It is believed to have been responsible for human death. Highly collectable but to be handled with great care if caught alive.

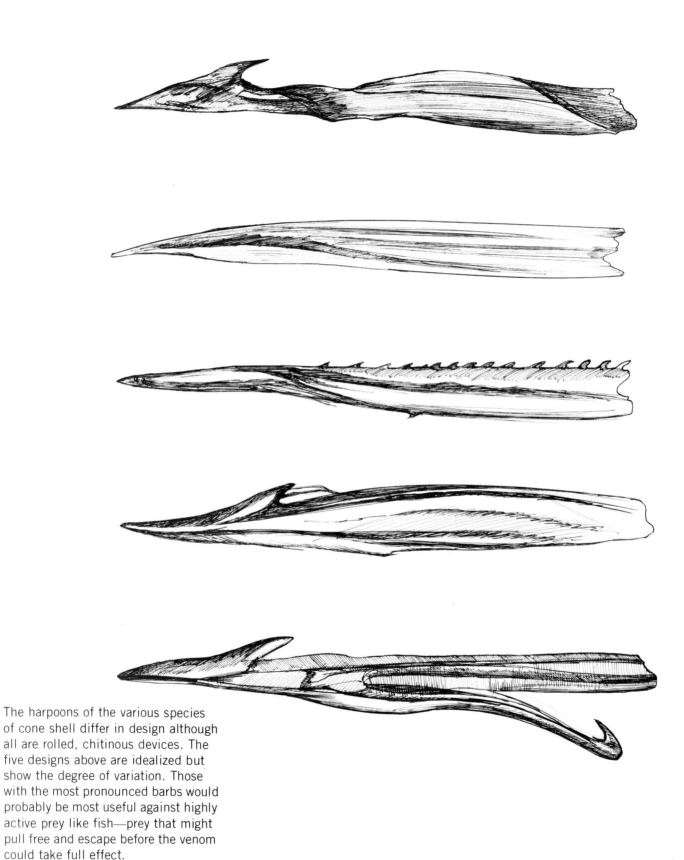

The harpoons of the various species of cone shell differ in design although all are rolled, chitinous devices. The five designs above are idealized but show the degree of variation. Those with the most pronounced barbs would probably be most useful against highly active prey like fish—prey that might pull free and escape before the venom could take full effect.

It was noted, though, that certain fish-eating cones, at least, jettison a small quantity of venom just before striking. It drifts away as a small and almost unnoticeable cloud. It was postulated that the venom in the water was used to distract or slow down the prey animal, giving the cone a chance to strike. It is much more likely, however, that what the cone is doing is clearing out the least potent venom from the front of the tube so that when the thick walls of the bulb contract, the harpoon will be charged with the best the animal has to offer. If this is so it would indicate that the harpoon is charged at the last moment, just prior to use. This practice could vary from species to species.

The pharynx where the forward end of the venom duct ends is a short section of the animal's forward gut. It is held in a ring of nerves and muscles. Also opening into this pharynx is the radula sac.

In the gastropods the *radula* is a kind of rasping organ, a very thin chitinous ribbon studded with row after row of teeth. These file-like teeth vary in number and form according to species. It is with this tooth-equipped ribbon that the carnivorous gastropod gains entry into the shells of prey animals. The teeth literally tear or wear away an opening. In the cones this radula has been drastically altered. The chitinous ribbon is missing altogether and the teeth have been freed and modified until they have become harpoons. It is a long evolutionary road from a solid, abrading tooth firmly anchored on a ribbon of such teeth and a lethal hypodermic dart—but the cones have successfully made the change.

The size and shape of the harpoons or darts vary, but in every case known they lie in two distinct bundles. The radula sac is asymmetrical, with two arms of different lengths. The walls are translucent when the organ is dissected out, and the harpoons can be seen lying in bundles in the two arms. The longer arm of the radula sac contains the larger bundle of darts. They are in various stages of development and lie with their sharp ends pointed toward the blind end of the sac. As they reach the end of their development period, they migrate or are forced over into the shorter arm of the sac. In the shorter arm lie a small bundle of fully formed harpoons ready for charging with venom and use. This short arm is aptly referred to as the quiver. The darts lie with points facing the opening into the pharynx. The darts themsevles are rolled tubes of a chitinous material, and each species has a characteristic design of barbs and serrations. The darts are glistening and transparent.

Each dart is attached to the wall of the radula sac by a ligament, but this apparently has nothing to do with supplying venom. There is no indication that venom is supplied to the harpoon after it is discharged. All of that is done beforehand.

The cone shells discharge their harpoons through a fleshy proboscis. It is actually a modified portion of the digestive tract and it extends forward from the pharynx probing ahead of the animal as it hunts. It has a doublewall, and is tubular. It is highly extensible and pulls in rather like a concertina when not in use. As it is brought into play it extends and smooths out. It is prehensile and can be used very effectively to hang onto organisms as slippery and active as fish. A tight ring of muscles surrounds the proboscis near its tip, and it is in this collar that the next harpoon is held.

The method by which the proboscis is charged with a loaded harpoon is not clear. It may turn inside out like the finger of a glove and reach back in to grasp the harpoon foremost in the short arm of the radula sac, or there may be another way the animal has of working the harpoon forward. One way or another a dart is charged with venom and brought forward for use.

As the cone hunts for prey a siphon extends out from the same end of the shell as the harpoon-equipped proboscis. As the creature crawls along it directs a steady stream of water back from the siphon to the *ctenidium*, or gill structure, and the *osphradium*, a chemosensory organ. It is with the *osphradium* that the shell apparently detects prey. By searching with the external siphon and evaluating with the sensor locked away in its shell, the animal can make the fine adjustments necessary to move in for the kill. Gliding along on or just below the surface of the sand, the cone shell closes in on its prey. Probing with its proboscis it locates exactly the right spot and pumps the harpoon forward with its self-contained charge of venom. The action is very, very quick and even a fish may be unable to move out of the way in time. Some cones seem to shoot their harpoons and others seem to stab theirs in with a pumping motion of the proboscis. At times more than one dart will be shot; at least one captive specimen was seen to utilize six for a single kill.

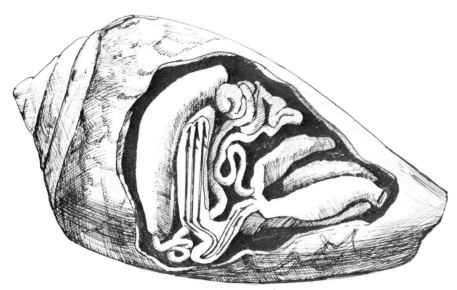

Diagrammatic dissection of a typical cone shell showing, in somewhat exaggerated proportions, the placement of the envenomating apparatus. The movement of the animal is toward the right, of course, and it is at that end the proboscis appears.

When a fish is harpooned it is skewered, and the dart is held by the ligament running from its posterior end back down the proboscis to the wall of the radula sac. A ring of muscle at the forward end of the proboscis tightens on the rear end of the harpoon and also helps hold against the fish's struggles. If the fish pulls free, as sometimes happens, the dart is discarded and another is brought forward. A dart is not used a second time.

Once the prey animal is stilled the cone glides over it and engulfs it with its highly extendible mouth. Very often it is necessary for the cone to leave its mouth outside dissolving prey too large to pull back in through the shell's inflexible aperture. At times like these the cone itself is highly vulnerable to predation. As was indicated earlier, a cone will use its harpoons defensively against creatures like man and octopus but not against other cones. It is not clear whether all predatory cones are immune to the venoms of prey species. These relationships have not been worked out.

The venom of the cone shell is a very complex substance, like all venoms. Physically it is granular in structure, or at least it so appears, and varies in color from white to black. Preliminary investigation has suggested that there are proteins (undoubtedly the dangerous fractions), quarternary ammonium compounds, and possibly some amines present. Without doubt there are other constituents that are not even suspected at this point. The venom is amazingly stable and withstands long freezing at very low temperatures and treatment with a variety of chemicals. It is a stable, potent, and dangerous substance.

Interestingly enough, we do not know specifically how the cone shell venom is dangerous. We do not understand, except in a superficial way, how this venom acts on, let us say, the human organism, although in vertebrate victims the muscles are acted on directly. There are indications that the venom interferes with neuromuscular transmission. The possibility that there is some action on the central nervous system should not be ruled out. The venom does not seem to affect individual nerves (although there is sharp disagreement on this) and some investigators say it *does not* block the junction between nerve and muscle. This would support the idea that the muscle itself is attacked by some substance which interferes with the contractile mechanism. If this is so the venom of the cone shell could be of great importance to man. Anything that can lock up, cripple, or otherwise inhibit a system within an organism could be the key to understanding it. The venom of the cone shell could be an extremely valuable research tool. But, for the time being, the actual mode of action of the venom escapes us. It will be valuable information for us to have.

The statistics of cone shell envenomation are not impressive. There are dangers, of course, to anyone handling them. This danger is apparently greatest in the Indo-West Pacific region. That is where the greatest variety and the greatest number of cones are found. Since the middle of the last century we know of 51 people stung by cones, and 15 of these cases were fatal, according to some tabulations.

Information available indicates that the geographer cone (*C. geographus*) is the most significantly venomous. At least more people seem to have been stung and killed by this cone than by any other species. Using Halstead and several other sources, the following list was compiled. It may contain some repetitions and certainly some cases are missing, but it is accurate enough to tell us the story we have to know.

The geographer cone, *Conus geographus*, a species believed capable of causing death in human beings. These highly collectable species are handled by more and more collectors and skin divers each year, and accidents are likely to occur.

CASES OF CONE SHELL ENVENOMATION

Conus aulicus (Court Cone)

Number	Locale	Date	Fatal
1	Moluccas	1848	No
1	New Guinea	1945(?)	No

Conus geographus (Geographer Cone)

Number	Locale	Date	Fatal
1	Japan	1889(?)	Yes
1	New Britain	1884(?)	No
2	Fiji Islands	1901	1 Yes
1	Okinawa	1927	No
1	Seychelles	1932	No
"Several"	Okinawa	1935	"Several"
1	Australia	1936	Yes
1	Seychelles	1946	No
1	New Caledonia	1963	Yes
1	Guam	1967	Yes

Number	Locale	Date	Fatal
	Conus imperialis (Imperial Cone)		
1	Seychelles	1957	No
	Conus lividus (Bluish Cone)		
1	Guam	1957	No
	Conus marmoreus (Marbled Cone)		
"Many"	Loyalty Islands	1877(?)	No
1	New Hebrides	1936(?)	No
	Conus nanus		
1	Oahu (Hawaii)	1955	No
	Conus obscurus		
1	Hawaii (island)	1947	No
1	Oahu (Hawaii)	1956	No
	Conus omaria (Pearled Cone)		
1	New Guinea	1954	No
	Conus pulicarius (Flea Cone)		
1	Oahu (Hawaii)	1955	No
	Conus striatus (Striated Cone)		
1	Australia	1935	Yes
	*Conus textile** (Textile Cone)		
1	Banda (Indonesia)	1705	Yes
5	Melanesia		2 Yes
	New Hebrides (2)	1859	3 No
	New Caledonia (1)	1874(?)	
	Loyalty Islands (1)	1936(?)	
	? (1)	?	
1	Australia	1935	Yes
1	Oahua (Hawaii)	1956	No
	Conus tulipa (Tulip Cone)		
1	Paumoto Island	1878(?)	No
1	New Caledonia	?	No
1	Tuamotu	?	No
1	Kwajalein	1954	No
	Conus sp. (Unidentified)		
"Many"	New Britain	1935	1 Yes
1	Hope Island (Australia)	1948	No

*Identity has been questioned in cases of human deaths.

When one considers the number of people exposed to cone shells every day on islands and atolls throughout the tropical oceans and then reviews the preceding list, it becomes clear that they do not represent much of a health menace *statistically*. On an individual basis, however, they demand respect and caution.

As far as this author knows there has never been a case of cone shell envenomation by someone not handling the shell. In other words, one must pick up a cone shell to become involved. The answer would seem simple. Avoid handling cone shells and you will not be envenomated. Things are never that easy, however. There are too many skin divers, too many shell collectors, too many people jetting to the most remote regions of the world for the danger of cone shell stingings to be so easily dismissed. More practical advice is required.

The harpoon equipped proboscis emerges from the shell's aperture about one-fifth of a shell-length from the base or pointed end. Most species apparently cannot whip their proboscis around past the angle of the shoulder at the wide end of the shell. However, some can, including the possibly formidable textile cone. Any shell that must be handled should be held by the wide end, base down, away from the body *in a gloved hand*. If a shell is found on or partially buried in the sand, a stick or other device should be used to rock it back and forth. This will cause almost any mollusc to retract into its shell at least temporarily.

After capture the cones should not be handled with ungloved hands and should *never* be allowed to crawl across the palm of the hand. Shells should never be retrieved from a collector's bag with an ungloved hand. Cone shells can be very active under such conditions. Of course, the shell, once the animal has been removed, is completely harmless.

In summary, the cone shells, all estimated 500 species of genus *Conus*, are active, aggressive molluscs equipped with a complicated but highly efficient apparatus for administering quick death or paralysis to fast-moving prey. The cones are not loathe to use this food-getting device defensively. It is on occasion deadly to man. Because of their great beauty and variety they are highly "collectable" shell specimens and are handled by the tens of thousands every year. There is a distinct measure of hazard in this fact.

OCTOPI

The second group of molluscs known to be venomous are the members of the class Cephalopoda, the head-footed molluscs. Human deaths are also attributable to this group, too. At least one small octopus that we shall be discussing at length has been said to have "the world's most potent poison." It is difficult to substantiate that kind of claim and, indeed, one must wonder how a comparison could be made. Without belaboring the point it is enough to say that at least one rather small octopus and possibly several are *among* the most seriously venomous animals on earth. There is no antivenin yet available.

The venomous secretions of certain cephaloped glands are *a* means or perhaps *the* means by which the octopus is able to obtain prey. Some species have been seen to drift over a crab and squirt venom into the water surrounding it, then drift away and wait. Within a minute or two the crab was seen to go into a period of great excitement and gradually lose control of its movements. Paralysis quickly followed and the octopus would then attack. The same species of octopus was seen to attack crabs without the preliminary stunning when they had been kept from food for several days. At least two researchers came to the conclusion that a less hungry octopus first quiets its victim, a more hungry specimen attacks outright.

It is quite likely that all species of octopi produce a venom of some kind, although this has not yet been proven. That there is a marked difference in octopus venoms and their toxicity cannot be denied. Within a single species it is believed that young, small, and unhealthy specimens are not as seriously venomous as large healthy ones. Females do not eat once they are caring for their bundles of fertilized eggs, and they will not even bite under these circumstances. (Once the female has a clump of fertilized eggs to aerate with sea water, she never eats again and dies soon after her eggs hatch.)

The venom produced by cephalopods has not been broken down into its toxic constituents. At least three substances, one of which is called *cephalotoxin*, are believed to be involved. There may be others. The components which researchers have succeeded in isolating are low in molecular weight and appear to be nonantigenic—they do not spur the development of antibodies. It is likely that hyaluronidase is present, for this enzyme has been found in the octopus's posterior salivary gland. Hyaluronidase is what is known as a spreading factor and helps the toxic elements of the venom to rapidly overwhelm prey.

The venom of the octopus creates a curare-like effect. It is a potent neurotoxin and is apparently secreted principally by the animal's posterior salivary glands. These glands are located above the brain some distance from the mouth. Ducts lead down from each of the two glands and join to form a common duct which then passes down through the brain to open into the buccal mass. The buccal mass is a pharyngeal cavity located above the mouth. It has thick muscular walls and is hidden behind the heavy muscular base of the animal's eight arms. The octopus's powerful jaws, shaped somewhat like a parrot's beak, are within the pharynx. There is no venom sac as such and the venom is apparently "spit" into the wound made by the sharp, powerful beak.

The beak of the octopus is not equipped with an envenomating apparatus, but in most species of octopus, the wound made by the beak is infused with venom. It is with this beak that the mollusc dismembers its prey.

The precise pharmacology of octopus venom is not fully understood. It apparently affects both the conductivity of nerves and the junctions where nerve and muscle meet. Death results from respiratory failure which in turn is a result of the failure of all voluntary muscular activity including that of the diaphragm.

The strange nature of the octopus salivary gland and its power to produce a dangerous substance has been recognized for over a century, but relatively little data about the effects on man was available until recently. There was an incident in 1945 where a film technician was bitten and had a relatively mild reaction, but there have been far more dramatic cases since. In 1950 there was a "probable" case where a person suffered curare-like envenomation, but it was not proven that an octopus was responsible. In 1954 a very highly publicized encounter took place, however, and there is no doubt about the general identity of the offending animal.

In September of 1954 a twenty-one-year-old man in apparent good health was bitten on the back of his shoulders by a small octopus. It had been handed to him by his spear-fishing companion and he had allowed the little animal to crawl across his bare back. Within minutes he showed signs of envenomation, within two hours he was dead in a hospital just four miles away. The victim was Kirke Dyson-Holland, the friend who later identified the octopus was John Baylis, but the exact species of the animal has been less clear. It has been variously identified as *Octopus rugosus*, *Octopus lunulata*, *Octopus maculosa*, and *Hapalochlaena maculosa*. The latter two animals are really one and the same; the difference occurs not in the animal but in the generic name assigned to it. Experts still do not agree and either name will do. It was apparently not *O. rugosus* and, although *O. lunulata* is easy to confuse with *O.* or *H. maculosa*, it was apparently not that animal either. What is important is that this is probably the first recorded case of a person killed by the venom of an octopus. The speed with which death came attracted a great deal of attention in medical circles.

There were two more cases in 1961 where octopi were positively identified. A young boy was bitten and made seriously ill but he survived. And a young man in his early twenties became seriously ill within three minutes after an encounter. He was handling an octopus (believed to be the same species that killed Kirke Dyson-Holland) and felt a small bite. Within minutes he experienced a feeling of numbness around his mouth followed by a similar sensation on his neck and the back of his head. He had difficulty in breathing, a feeling of weakness in his limbs, and a burning sensation on his skin whenever he was touched. Total paralysis set in quickly, with his eye muscles being the last to fail him. He was kept alive by artificial respiration and a tube inserted in his throat. Within a few days he experienced total recovery. Why did he survive and Kirke Dyson-Holland die? They were both healthy young men, both twenty-one years old, both apparently athletic. Dyson-Holland had had some history of asthma, some allergic background. This second young man had no such medical history. That *could have been* a factor. It has not been clearly identified as such.

In July 1962 there was a fifth case in Australia. *O. maculosa* was identified as the offending animal. The victim was a healthy thirty-three-year-old man and again total paralysis and complete recovery were reported. It began to seem as if the unfortunate Kirke Dyson-Holland had a peculiar sensitivity to the venom. It seemed most unusual that there should be only the one death. The total paralysis experienced by several other victims would certainly indicate that a dangerous substance was involved. However, in June 1967 there was another death.

The deadly blue-ringed octopus (*Hapalochlaena maculosa*) in a rock pool, Queensland, Australia. This is the most seriously venomous octopus presently known or identified as such. Its bite is deadly to man in at least some cases.

On June 21, 1967, Private James Arthur Ward was exploring the rock pools in the Camp Cove area near Sydney with two other soldiers. He was twenty-three years old and in excellent health as evidenced by the fact that he had been inducted into the Australian Army two days earlier. Early in the afternoon the soldiers encountered a small specimen of *O. maculosa*. According to the testimony of the other men it was less than four inches across the arms. Private Ward was dead of respiratory failure in less than 90 minutes.

The octopus that is known to have killed Ward, is believed to have killed Dyson-Holland, and is known to have made several other people seriously ill* is generally referred to as the blue-ringed octopus. When at rest the animal has dark brown or ochre bands over its body and arms. Blue circles are superimposed on these bands. When the animal is angry (a mildly anthropomorphic interpretation of an octopus's mood) or otherwise disturbed, the ground colors darken dramatically and the blue rings seem to light up until they become an iridescent, almost peacock blue.

The species is widespread in the coastal waters of Australia and is commonly encountered in crevices, rock pools, and submarine caves. It has been found to a depth of 30 plus feet. The species is reportedly very common in Lake Macquarie, a salt water lake north of Sydney.

There is no doubt that the envenomizing system of the octopus is primarily a food-getting apparatus. Some species are apparently more prone to utilize their venom for defense than others. It is apparent that some species are very much more toxic than others. Until more is known it would seem advisable to handle all octopi, especially small ones from the western Pacific Ocean, with extreme care. They should be grasped, when it is necessary to handle them, around the body *above* the arms and should not be allowed to bring their underside and especially their mouth into contact with bare skin. They do not have a fang system of any kind, and they do not have the cone shell's system of harpoons. To envenomate they must first make a wound with their beak and then secrete or spit venom brought down from their salivary glands into the wound. Unlike a snake's fangs or a cone's harpoons, however, the beak of the octopus has other uses. The beak is what the octopus eats with, the device it uses to open and render its prey. It apparently serves its function in the envenomating system well enough for no other device ever to have evolved in these ancient and endlessly fascinating animals.**

*There have been other cases of envenomation, but these are not detailed here since they do not differ substantially from those noted.

** Some octopi, as we have seen, use the stinging tentacles of jellyfish to defend themselves. However, this is a tool-using adaptation and although startling in its implications not a subject we would be justified in elaborating upon here.

Chapter 4

THE

ARACHNIDS

The Brazilian spider *Aranha* in the warning pose. The biting mechanism can be clearly seen on the underside of the head. The toxicity of these animals is generally grossly exaggerated although some species can cause serious systemic reactions.

he phylum Arthropoda is the largest and one of the most diverse phyla in the animal kingdom. It includes fully 80 percent of the known animal forms. It contains, by popular names, the horseshoe crabs, arachnids (spiders, scorpions, and their kin), the crustaceans (shrimp, true crabs, lobsters, and various terrestrial "bugs"), the millipedes, centipedes, and the almost incomprehensibly vast world of insects. In total, there are probably upwards of a million and a half arthropod species although many have not yet been adequately described. Some of the most fascinating of these are the subject of this chapter—class Arachnida—the spiders and scorpions. We will also touch on two other classes.

SPIDERS

Spiders do not have mandibles or jaws and are unable to chew. They are, therefore, fluid feeders. In place of jaws they have appendages called *chelicerae* that are used for cutting or piercing. Once they have penetrated the body walls of their prey they suck up the fluids and, in many cases, digest the solid tissues by liquefying them with a digestive enzyme. The venoms with which spiders kill their prey are also the chemical agents they use in their own defense.

GARDNER WEBB COLLEGE LIBRARY

Most spiders are venomous. They are equipped with a pair of venom glands each one of which feeds a fang-like structure, the chelicera, through efferent ducts. However, even most of the truly venomous spiders are harmless to man. The venom they are capable of injecting is either so feeble or the quantity so small (or both) or the fangs so weak that no harm can be done. This fact is stated in direct contradiction to popular belief. Most people are apparently convinced that all spiders are deadly dangerous under all conditions. This kind of an attitude is partially understandable when directed toward snakes, in areas where at least some snakes are seriously venomous. It is strange, though, that it should be directed toward spiders even in areas where no spiders exist that could be dangerous to any large animal.

Of the approximately 30,000 species of spiders only a few can be dangerous to man. They belong, almost without exception, to a few genera:

KNOWN DANGEROUS SPIDER GENERA

Capable of causing human death

Atrax	*Harpactirella*
Latrodectus	*Loxosceles*
Phoneutria	*Trechona*

Possibly capable of causing human death

Chiracanthium	*Lithyphantes*
Mastophora	

Two forms of spider chelicerae. These envenomating devices vary greatly from species to species, the form accommodating the type of motion for which the animal is designed. Some jab down, some move laterally, and in each case the design of the tip of the chelicerae provides maximum opportunity for penetration and envenomation.

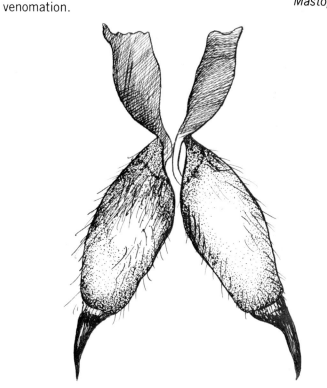
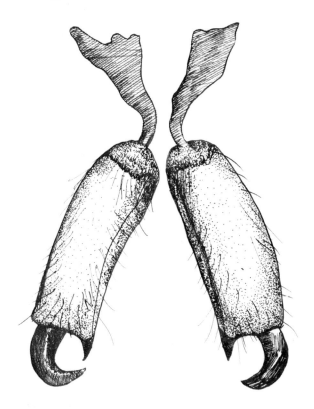

In addition to causing death, or *instead* of causing death in most cases, spiders can cause localized injuries. Bücherl (1971) designates certain genera as causing "superficial wounds" and others as causing "deeper and larger wounds." He suggests the following:

Localized Superficial Wounds	Deeper, Larger Wounds
Araneus	*Acanthoscurria*
Argiope	*Avicularia*
Dendryphantes	*Loxosceles*
Lycosa	*Megaphobema*
Nephila	*Pamphobeteus*
Tegenaria	*Phormictopus*
	Theraphosa
	Xenesthis

It will be noted that as we discuss the spiders, except in a few cases, we will discuss entire genera rather than individual forms. This is necessary because of the problem of dealing with 30,000 species. It should be noted that not all species in a genus will be of the same virulence. Thus, when we speak of a genus it is to say that *some* species in that genus exhibit a characteristic or degree of toxicity.

The envenomating aparatus of spiders vary greatly. In some spiders the *chelicerae* point downward and the movement is largely vertical. In others the fang-like devices are diaxial and movement is horizontal. The size, shape, and musculature of the venom glands, the internal structure and their ducts also vary according to species, genus, or family.

Spiders vary greatly in life-style and the ways in which they use their venoms are affected by this fact. Some spiders (like the deadly black widow, *Latrodectus mactans*) are web spinners, nearly blind animals that spend their lives in and about their web. Some are semi-vagrant and move about rather more than the web spinners. Others are truly vagrant and actively prowl for food, while still others are burrowers and spend a great deal of time underground. Some are semi-aquatic.

All spiders belong to the order Araneida (sometimes given as Araneae) which, in turn, is usually divided into two suborders: Orthognatha (the so-called "tarantulas") and Labidognatha (the so-called "true spiders"). In both groups the device that actually introduces the venom is formed of two segments—the basal and the fang or claw-like *chelicerae*. The device is curving and hollow, with the channel closer to the convex side. The opening is at the tip. The fang part (we will refer to it as a *fang* from here on) may be anywhere from 0.0125 to .495 inches long. Interestingly enough, the very deadly *Latrodectus* species have fangs toward the smaller end of the spectrum. The figures given are for the fang part alone and do not include the larger, basal section.

In the tarantula-like spiders the thin white duct that carries the venom forward is about the same length as the fang itself. In the so-called *true spiders* the duct is equal in length to the entire *chelicera,* both the fang part and the basal segment. In some of the species with longer ducts a small ampule-like structure is located at the forward end apparently to store venom for instant use. In some the duct itself is forced open by the pressure of freshly exuded venom (at least this is so in the case of the tarantula spiders). Special duct-opening muscles are found in some species with longer ducts.

The venom glands of the spiders are divisible into three separate parts. There is a muscular sheath around the outside, a basic holding or basement membrane, and the layers of venom-producing epithelial cells. The glands are always whitish in color. The surrounding muscles are gathered into bundles and the individual bundles as well as the mass as a whole are covered by a very fine sheath or sarcolemma. The orientation of the muscles

Not all spiders are limited to insect prey. Some are large enough and toxic enough to subdue even reptilian prey. Here a Honduran tarantula eats a tropical green-spotted racer.

to the gland is not in any fixed pattern, but appears random, though often spiral. There may be as many as 160 bundles of muscle fibers on one side of a single venom gland. The structure, obviously, is complex. Dimensions of the venom glands of possibly dangerous species range from (in inches) .47 x .06 and .40 x .106 (to indicate two different shapes) all the way down to .058 x .007. The amount of damage that can be done with the secretions of such extremely small devices is amazing.

In their experimental studies of spider venom (and this is true of all arachnid studies), scientists have had a problem obtaining pure venom for their tests. Snakes, for example, are simply milked to get pure venom. But obtaining pure venom from any animal as small as a spider is a more difficult task. A common method has been to remove the venom glands and grind them up into a uniform solution. Since only a small part of a gland will be venom (and that percentage could be extremely difficult to establish as a constant), the results of tests with such solutions could be misleading.

Another method employed is actually "milking" using electrical stimulation. It works with both spiders and scorpions although with different degrees of success. It is quite possible that the characteristics of the venom in the gland are different from those of the material once it leaves the gland. It is known, for example, that spider *silk* changes dramatically as a result of the mechanical stress of expulsion. It is not known whether or not this is true of their venom as well. It is also possible that when spiders are electrically "milked" sudden muscular contractions may force out fluids different from those voluntarily discharged during the envenomation of prey. The answers to these questions are not available.

A further method of obtaining venom is to mechanically irritate a spider until it *stings* a membrane or discharges its venom into a receptacle. It is slow, difficult work, but the venom obtained is probably pure and much the same as that discharged naturally. One way or another, scientists have had enough spider venom for study and testing. It is a remarkable substance.

When the venom of *Phoneutria nigriventer*, a common South American spider, is injected into experimental animals a series of violent systemic reactions are recorded. There is intense local pain, violent sneezing, the eyes water excessively, and there are abnormal amounts of salivation. Other reactions include a generalized weakness, abnormal pain and irregular heart beat, difficulty in breathing, drowsiness, vomiting, priapism and ejaculation, and blood in the feces. Some animals have died even from relatively small subcutaneous injections. When injected into the blood stream the venom proves to be very toxic and causes a sharp drop in blood pressure (Schenberg and Lima,

1971). The wide range of neurotoxic reactions noted indicate a number of different factors are involved. Some of these may have been isolated, but clearly a great deal of work must be done before we understand the chemistry or pharmacology of a substance capable of causing such violent reactions in such small dosages.

The genus *Latrodectus* contains some of the most seriously venomous spiders yet recorded. There are at least six species and they are universally feared for what they are—the most dangerous spiders of all. They are, of course, known by a great many vernacular names but these are the species currently recognized:

GENUS *LATRODECTUS**

Species	Range
L. geometricus	Cosmotropical
L. mactans mactans	Cosmotropical and temperate zones. Most wide-spread in the U.S. to Argentina.
L. mactans tredecimguttatus	Mediterranean region, Central Asia, Abyssinia, Arabia
L. mactans indistinctus	Abyssinia, East and South Africa
L. mactans menavodi	Madagascar
L. mactans hasselti	India to Australia and New Zealand
L. pallidus	Russia, Syria, Palestine, Iran, Libya, Tripolitania
L. curacaviensis	Southern Canada to Patagonia (absent in Mexico and Central America)
L. hystrix	Aden and Yemen
L. dahli	Iran, Sokotra

The symptoms produced by the bite of the black widow spider have been reported scores of times in both technical and popular literature. Their summation in a circular prepared by Merck Sharp & Dohme for their Antivenin is as concise as any:

> Local muscular cramps begin from 15 minutes to several hours after the bite which usually produces a sharp pain similar to that caused by puncture with a needle. The exact sequence of symptoms depends somewhat on the location of the bite. The venom acts on the myoneural junctions or on the nerve end-

*After Levi as reported by Bücherl, 1971. The taxonomy of this genus is in a state of chaos and our choice of any one expert is, in a sense, arbitrary. Since our task here is not arbitration, arbitrary we shall be.

ings, causing an ascending motor paralysis or destruction of the peripheral nerve endings. The groups of muscles most frequently affected at first are those of the thigh, shoulder, and back. After a varying length of time, the pain becomes more severe, spreading to the abdomen, and weakness and tremor usually develop. The abdominal muscles assume a boardlike rigidity, but tenderness is slight. Respiration is thoracic. The patient is restless and anxious. Feeble pulse, cold, clammy skin, labored breathing and speech, light stupor, and delirium may occur. Convulsions also may occur particularly in small children. The temperature may be normal or slightly elevated. Urinary retention, shock, cyanosis, nausea and vomiting, insomnia, and cold sweats also have been reported. . . .

In 1943, as a teen-age medical corpsman on bivouac with a national guard unit in Massachusetts, I had to help treat a young guardsman who had been bitten by a black widow spider. He had broken the rules and used an abandoned outdoor privy near where we were camped. A black widow spider bit him on the penis. It was some hours before an ambulance reached us. During that time the doctor feared for the boy's life. He suffered all the symptoms of extreme shock; and the terrible board-like rigidity of his abdomen, the agonizing "charley horses" in his arms, and legs, and his sobbing are not easy to forget. It was then that I think I first began to wonder about the incredible power of so small an organism. Why and how had it been equipped to so seriously impair the health of a 160-pound teen-age human being? It ate flies.

This grab bag of symptoms indicative of severe envenomation may continue to increase in severity for as much as a day following onset. Subsidence will be steady after that, but some symptoms such as general weakness, nervousness, and brief muscular spasms can continue for weeks and even months. In about 5 percent of the cases there is at least the possibility of death. As reported in a number of places, there were 1,291 known black widow envenomations in the United States during the 218-year period 1726 to 1943. Of that number 55 were fatal—a mortality rate of a little over 4 percent.

Bites by the brown widow (*L. geometricus*) are much less severe than those by the black widow, and have not resulted in any reported deaths. The symptoms are generally local but there can be nausea and vomiting and radiating pain. They are generally like the symptoms from black widow spider envenomation only very much less serious. Baerg (1959) reports on the bite by a female *L. bishopi,* the red-legged widow (species not recognized by Levi), and the symptoms resulting from this were milder yet.

A ''baboon spider'' in Kenya, East Africa, about to feed on a chameleon it has just killed with a venomous bite.

There seems to be a basic syndrome connected with *Latrodectus* envenomation, with the difference between species being a shocking difference in degree. Commenting on the *Latrodectus* neurotoxin, Horen (1963) stated that it affects "... chiefly the spinal cord, but also induces cytotoxic reactions. Generalized injury occurs in several organs in the form of parenchymatous necrosis involving blood vessels, epithelial cells, nervous tissue, and lymphnoid cells."

L. M. hasseltii, the red-back spider, is found from India to Australia and New Zealand. The Commonwealth Serum Laboratories in Melbourne produce an antivenin and the circular that accompanies it presents this review of red-back spider bite symptoms:

> The bite may cause a sharp sting or burning sensation, or it may not be felt. Soon afterwards redness and oedema develop at the site of the bite accompanied by localized pain and sweating. The pain increases in severity and spreads throughout the body. It is often burning in character. Profuse sweating usually occurs and may be accompanied by shivering, nausea, restlessness and muscular weakness. A variety of subjective symptoms may be experienced.

Again we see the generalized neurotoxic syndrome inflicted by a member of the widespread, locally common, universally dreaded genus *Latrodectus*. McCrone and Hatala (1967) propose the structure of *Latrodectus mactans* venom: seven protein and three nonprotein fractions. One of these fractions, labeled Protein Fraction B, is apparently the lethal factor for mammals. When isolated it has 20 times the lethality of whole venom.

Latrodectus venom is essentially neurotoxic and local symptoms are largely absent. This is not true of all spiders. Spiders of genus *Loxosceles* cause significant local damage. The brown recluse spider or fiddleback (*L. reclusa*) is a domestic species that is apparently expanding its range in the United States. Although it is also found in areas away from buildings, the brown recluse is characteristically found in and around man-made structures. It is common in garages, attics, and basements as well as in closets and storage rooms.

The brown recluse was probably restricted to the mid-South and Midwest until recently. Its habit of hiding in clothing, trunks, and boxes has resulted in its being spread across much of the United States on moving vans if in no other way. It is a long-lived spider, with a possible life span of several years, and each female produces up to 300 fertile eggs a year. Since the

The brown recluse spider, *Loxosceles reclusa.* Also known as the fiddleback, this midwestern North American species is having its range expanded by the activities of man. At least seven human deaths are attributable.

The seriously venomous Australian red-back spider (*Latrodectus hasseltii*) is venomous enough to prey on lizards and mice and to seriously affect the health of a human victim. It is sometimes given as a subspecies of the American black widow and sometimes accorded full species status.

female protects her young until they are at least partially grown, the survival rate is probably high) The female does not kill her mate the way the black widow sometimes does and the male of this species is as venomous as his mate. Neither sex is usually deadly to man but six fatal cases are on record.

(The bite itself is not particularly painful and its true nature can be obscured at first.) Onset of symptoms may be delayed as much as eight hours. (There may be discoloration, blistering, and some hemolysis. Ulceration will follow these first symptoms)

When the reaction to the *Loxosceles* venom is particularly strong there may be a generalized rash, fever, and nausea. Severe abdominal cramps may last for hours. The wound itself may not be particularly painful and may range from about the size of a dime to the size of a half dollar or larger. The skin in the area will die, rot, and eventually drop away. A depression not unlike a bullet wound in appearance may remain open and fussy for several months before a good protective scab forms. In 1968 a fifty-six-year-old man in Alabama died five weeks after being bitten by a brown recluse. The open wound on his leg was ten inches in diameter at the time of his death. In another case a victim survived three weeks and then died. These deaths, however, were extremely unusual. Most brown recluse bite victims experience local reactions only. The most distressing part of these is the prolonged ulceration.*

The matter of sexual dimorphism of venomous characteristics is open to debate. In his review of venomous spiders, Bücherl (1971) states: *"No male of any spider species is capable of injecting a deadly dose of venom into the human body."* In their discussion of the Sydney funnel web spider (*Atrax robustus*), Gilbo and Coles (1964) state: *"These workers [Wiener and Kaire] have shown the venom of both male and female spiders to be lethal to laboratory animals, but in the few cases where the sex of the spider inflicting bites has been determined, only male spiders have been definitely proved to have caused death in adult humans."*

The genus *Atrax*, may be unique in that the males are *more* seriously venomous than females. Between 1927 and 1961 there were at least ten human deaths attributable to males of the genus and more probably went undiagnosed or were improperly recorded. The venom contains digestive enzymes and "predigestion" by these fractions could account for severe local reactions of distressing duration. Their fangs are particularly formidable and, according to Saul Wiener (1957), they can penetrate the skull of a chicken.

---

*Minton points out (private communication) that *Loxosceles laeta* is a widespread species in South America. It accounts for some "horrible cases of necrotic arachnidism plus some deaths from hemolytic crisis." It has been introduced into the United States and may be established in Boston and Los Angeles.

Tests on the venoms of both sexes of *A. robustus* show that the females have larger venom glands, larger fangs, and more venom. Their venom is much less toxic than the males', however, perhaps with a factor of 6/1. It has been suggested that because the male has less venom and injects it into a shallower wound it needs a stronger venom. On the surface this seems logical, but why is it not true of other species as well? This characteristic may be, as we said above, unique to this species. (It has also been suggested that the *Latrodectus* spiders, particularly *L. mactans,* the black widow, are so much more venomous than other spiders because they prey on hard, tough beetles. That, too, has all the earmarks of oversimplification. The reasons for the varying levels of toxicity in venomous animals remain a mystery. If there were a rule of thumb that aligned degree of toxicity to type of prey, most of the venomous animals of the world would be guilty of overkill. Certainly no animal the size of a spider, however big or ferocious the beetle it had to face, would need a venom that can flatten a 200-pound man and keep him ill for days and perhaps weeks, and in some cases even result in his death.* I know of no rule, no hint of a rule, that applies in this strange matter. If we are ever able to determine a guide that adjusts our concept of toxicity with our concept of need, we will know a great deal more about wild animals than we do now.)

A. robustus, the Sydney funnel web, is not the only member of *Atrax* dangerous to mammals. Five members of the genus in Australia are believed capable of inflicting serious intoxication. Irwin (1952) reports on a case involving *A. formidabilis,* the North Coast funnel web. A seven-year-old boy was bitten on the arm at 9:30 in the morning while playing on his family's kitchen floor. A large spider was captured and later identified as *A. formidabilis.* (The funnel webs are large animals. A fully grown female may be an inch and a half long, a male just slightly over an inch.) The boy was hospitalized and for the remainder of the day his condition was good, but he was kept over night. By the time the doctor arrived at the hospital the following morning the boy was near death. His skin was cold and clammy and muddy grey in color. His lips were blue and his temperature was so far below normal the standard clinical thermometer could not record it. His pulse was weak and thin and was recorded at 180 beats per minute. He was breathing at the rate of 36 times a minute and there were bubbling sounds in his chest. His parents were advised that there was little hope, but radical treatment was initiated and the boy did survive. At the end of four days he was released from the hospital. Interestingly enough, tiger snake antivenin was administered. The tiger snake's venom is powerfully neurotoxic (as we will be seeing in the chapter covering Australian elapids) and

*Unless one accepts the "accidental" theory.

so is the venom of the *Atrax* spider. What role the antivenin actually played in the recovery of the boy cannot be determined.* The attitude of the doctor administering the medication was that it could not hurt the boy and just might offer some relief from the powerful neurotoxic syndrome from which he was suffering.

Spiders are not wholly inconsequential from the public health point of view. Between 1957 and 1960 they killed a known 21 people in the United States alone. (In that same period of time bees and wasps killed 127 and snakes killed 57.) When one considers, however, the number of spiders we are constantly being exposed to it is clear how reluctant most of them must be to bite. One British scientist has estimated that in a single acre of farm land there may be close to two and a quarter *million* spiders. Most of them, by far, will be harmless except to insects and other small animals on which they feed. Those few with the capacity to envenomate anything as large as a human being apparently temper their toxin with a rather benign disposition.

If the smaller spiders are capable of causing dread in some people, the giant and hairy so-called tarantulas apparently cause stark terror. Unlike the case with some of the smaller shiny spiders, the fear is unwarranted. Baerg (1958) reviewed his 30 years' work with spiders and concluded that no North American tarantula is dangerously venomous to man. A few, when induced to bite, cause a pinprick sensation, some a local reaction that can include pain and swelling, but none exhibit significant systemic effects. There are, however, over 300 species and tests are not complete or conclusive. Maretic (1967) reported on the venom of an East African orthognath spider, genus *Pterinochilus*. He found the venom to have serious effects on laboratory animals, including mice and guinea pigs, and concluded that the spider is seriously venomous. He observed that very little is known about the venoms of African and Asian species of this group. That is, in fact, quite true. Relatively little is known about the venoms of most species. Tests on laboratory animals are only partially revealing. Different species of mammals show different tolerances and this would necessarily be true of mice and men in at least some cases. In instances where rather elaborate conclusions are drawn from a single clinical report, there must be at least the suspicion that all of the answers are not available. Without doubt allergic reactions enter into the picture. In some cases, highly toxic species may not inject venom in sufficient quantity to exact anything but the most trivial reaction. (This is true of snakes, certainly, and records shown that as many as a quarter of the bites by rattlesnakes may be "dry.")

*It was probably minimal. Recovery, in fact, may have been unrelated to what was really "nonspecific protein therapy."

As indicated at the beginning of the chapter, only a very cursory review of the venomous spiders is provided here. With between 25,000 and 30,000 species already known (and perhaps that many again yet to be described) and with anything like conclusive evidence of toxicity available for only a very small percentage, a definitive review is not even possible at this time. There have been about a thousand papers published to date on venomous spiders, but like this chapter many are little more than reviews of information revealed in earlier reports. Much, much more work needs to be done before the possible measure of the spider-bite problem can be ascertained. It is clear, though, that spiders as a group are much more helpful than harmful. They

The "hairy" tarantula of the American Southwest, a much maligned although virtually harmless spider. These animals are venomous, of course, but they are not especially aggressive and their toxin is mild. Their bite has been compared with a beesting.

exist in such staggering numbers, even in relatively restricted areas, that the quantity of insects they consume cannot help but be significant. Certainly, spiders consume billions upon billions of insects every day. That has to be viewed as a service to mankind.

Adapted from Bücherl (1971) and others, the following list includes spider genera and species known to be dangerous, thought by some to be dangerous, and probably not dangerous although accused, region by region.

Region	Proven Dangerous	Not Yet Proven Dangerous	Accused But Dubious
Australia and New Zealand	*Atrax* *Latrodectus m. hasselti*		
Pacific (Hawaii)	*Latrodectus m. Mactans*	*Chiracanthium**	
Africa	*Harpactirella* *Latrodectus m. tredecimguttatus* *L.m. cinctus* *L.m. menavodi* *L.m. pallidus*		
South and Central America	*Loxosceles* *Latrodectus m. mactans* *L. curacaviensis* *Lycosa* *Phoneutria*	*Trechona* *Xenesthis* *Megaphobema* *Pamphobeteus* *Lasiodora* *Theraphosa* *Phormictopus* *Acanthoscurria* *Lithyphantes* *Mastophora* *Chiracanthium* *Dendryphantes*	*Heteropoda venatoria* *Sericopelma* *Eurypelma* *Segestria* *Filistata*
North America	*Latrodectus m. mactans* *Latrodectus geometricu* *Latrodectus curacaviensis* *Loxosceles reclusa* *Loxosceles unicolor*	*Chiracanthium inclusum** *C. diversum**	*Eurypelma*
Mediterranean Region, Asia Minor and Eastern Europe	*Latrodectus m. tredecimguttatus* *L. pallidus* *Loxosceles rufescens*		
India	*Latrodectus m. hasselti*		

*It seems certain that at least some species of the genus *Chiracanthium* are dangerous.

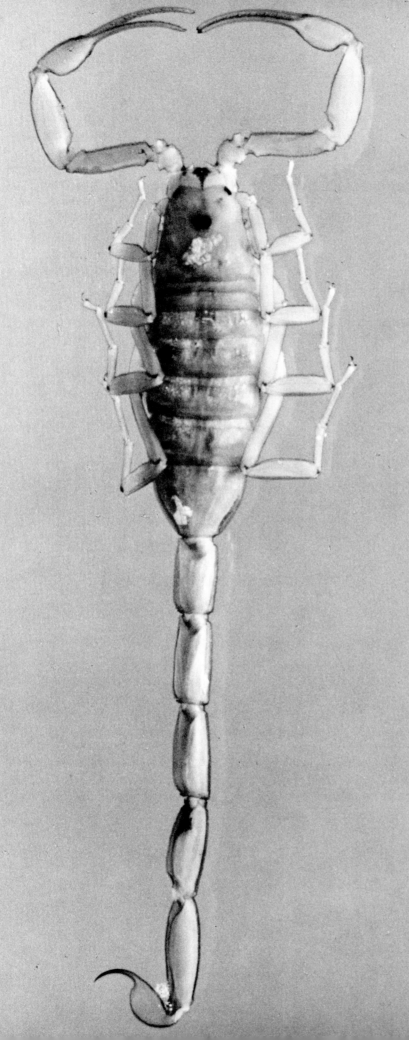

A female specimen of the scorpion *Centruroides sculpuratus*. This species, from the American Southwest, has caused human deaths. The genus is the most dangerous in North America.

SCORPIONS

The scorpions constitute only a small group in the class Arachnida. About 650 species are recorded although there will certainly be more recognized in the future. People often think of scorpions as strictly tropical animals. This is a misconception. They are found in the Pacific Northwest, in the Swiss Alps, in Montana, and a number of other areas well outside of the tropics.

Scorpions range from about three quarters of an inch to over eight inches in length. Size and toxicity do not seem to correlate except by accident. Some of the most deadly are among the smaller known species.

As for the potentially deadly nature of some species, the statistics speak for themselves. Between 1929 and 1948, 94 people died in the state of Arizona from bites and stings of venomous animals. Of the 94, 68 percent, or 64 people, died as a result of stings by scorpions. (Arizona, as we shall be seeing, is the apparent epicenter of rattlesnake development. It also has a coral snake, the Gila monster, and the black widow spider. Clearly, scorpions are a problem in some areas.)

Scorpions are interesting animals. They may have anywhere from two to 12 eyes, but they are virtually blind. They can differentiate between light and dark, but they could not, for instance, "see" a man. They manage to get around as efficiently as they do with the use of *pectines,* feelers located just to the rear of the last pair of their eight legs. Touch is very important to them.

Like the spiders the scorpions have chelicerae, but unlike the spiders these are not equipped with venom glands. They are simply pinchers. In this case the venom is located at the other end of the animal.

What people often refer to as the scorpion's tail really constitutes part of its abdomen. The trunk is comprised of eight segments, a broad one (cephalothorax) and seven narrow ones. The seven narrow segments combined with the first five segments of the so-called *tail* constitute the animal's abdomen. At the very end, attached to the last small portion of the abdomen, is the stinging apparatus. It is separate and apart from the animal's other bodily functions and can be removed without interfering with them.

In that bulbous final segment, or *telson,* are two venom glands. They lead to a single stinger and each has its own opening near the tip. The narrower after segments of the abdomen (the "tail") have what amount to almost universal joints, and the stinger can be quickly maneuvered into position. Although a scorpion may be slow to react because of being nearly blind, once it has determined to sting it is fast and accurate. Typically, an alarmed scorpion carries its rear abdomenal segments arched up over the rest of its body with its stinger at the ready. A short, abrupt jab forward and the act of envenomation is accomplished.

The scorpions, the world's oldest surviving land animals, prey on a number of forms. They eat other scorpions (including young of their own kind), spiders (including black widows), various beetles, flies, grubs, roaches, grasshoppers, centipedes, and worms. They seek prey by feel and not by sight and are almost exclusively nocturnal. When possible they subdue their prey mechanically, but if the struggling gets out of hand (they rip their prey apart to eat it) they use their venom. This is used as a last resort, for it takes the animals an appreciable length of time to recharge their venom glands. It may require as long as several weeks following a complete discharge.

Magnified view of the scorpion's venom apparatus. The "tail" or rear abdominal sections can be whipped forward with great speed enabling the two venom glands to discharge their loads through paired ducts and exit lumens.

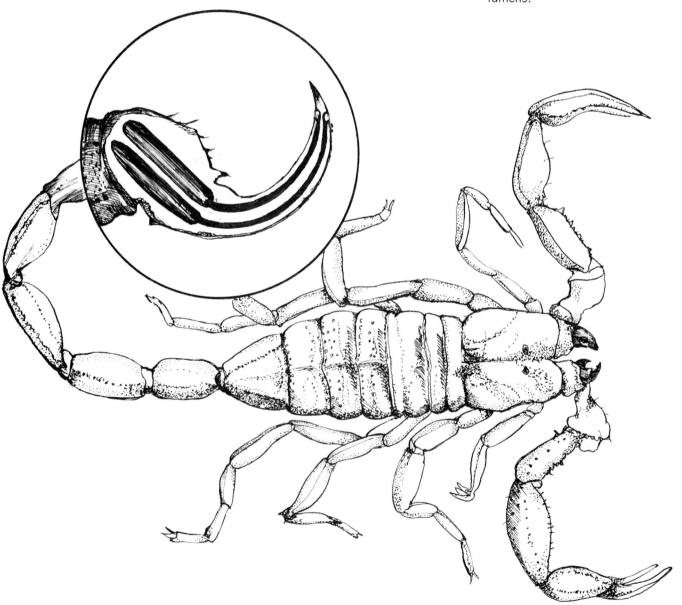

The external skeleton of the scorpion is called the cuticle. It is firm and against it, on the inside of the telson, the two venom glands are clamped by a wall of muscles. When the muscles contract the venom glands are squeezed against the cuticle and the contents are forced out through small orifices.

Scorpion venom is complex and has been subjected to a great deal of study and experimentation. It is largely protein in nature and may contain a number of lethal factors. Some of the venoms include enzymes, but not many. There are also spreading factors (hyaluronidase) in the venoms of some species. There are a number of variables and these account for the fact that some scorpions are deadly to organisms the size of man while others are harmless.

In the seriously venomous species the Scorpion sting can be extremely painful. A burning sensation may give way to a strange prickling accompanied by goose flesh and bristling hair. (This latter symptom has the interesting technical name of *horripilation*.) The bristling and tingling can radiate out from the site of the wound and become quite intense. It, in turn, can be followed by a feeling of insensitivity.

General symptoms vary in their time of onset. Some may be rapid, others may be slow in appearing. There may be agitation bordering on violence. Even sedatives may be unsuccessful in quieting the sick person.* There may be muscular cramps in the limbs, in the abdomen (this resembles the reaction to many spider bites), and even in the pharynx. Convulsions may occur in children. Priapism may occur (again we see a similarity to reactions to some spider bites) and tendon reflexes are reduced or vanish completely. Corneal reflex is diminished or disappears. The victim usually remains conscious during these stages and experiences a sensation of depression and even deep anguish. There is excessive salivation and perspiration. The eyes water and sight is impaired. The eyes may bulge and the pulse rate increases sharply and becomes, in time, erratic. Heart muscle damage may occur. The blood pressure may soar at first, but then it begins to drop and will be unstable. The temperature, too, lacks stability and goes up and down. Respiration is irregular and it can appear as if the whole body and all of its functions were out of synchronization. Death, when it occurs, is caused by respiratory paralysis. Scorpion sting can cause acute pancreatitis. There may be sneezing fits and, interestingly enough, people who handle scorpions in the laboratory report a sensitivity that reflects itself in long bouts of sneezing.

The first reliable sign that the victim is going to suffer profound distress is vomiting. Blood may be present and the stress

*Demerol and morphine are contra-indicated and their use in a scorpion sting victim could prove lethal.

on the system is reflected in diarrhea. When these two symptoms occur the prognosis deteriorates. Death may occur after a few minutes or not for as long as 30 hours. A very important characteristic of scorpion envenomation is relapse. Many cases have been reported in which all signs of distress cleared up and the patient appeared to be either fully recovered or well on the road to his former state of health when a total relapse occurred. In one series (discussed by Balozet, 1971) 13 out of a total of 17 relapses ended in death.

Deoras (1961) relates the case of a teen-age girl who was stung by an Indian scorpion of the genus *Palamneus*. She was first seen at the hospital at 10:30 A.M., an hour and a half after being stung. She was in a state of collapse, pulseless, and perspiring profusely. Her skin was cold and clammy and she was having chills. She had vomited a number of times after the encounter and was complaining of a burning sensation in her abdoman. She was medicated with coramine, glucose, and calcium gluconate and felt better almost immediately. Her pulse strengthened and could be recorded. But, she then began to shiver again and her pulse once again became imperceptible. She was medicated with adrenaline and an antihistaminic, but the perspiring and shivering continued. When the shivering eased again she was given hypertonic saline. Oxygen was given intranasally from the time of admission. She remained mentally alert and was able to respond to questions. She continued to perspire, however, and exhibited difficulty in breathing. At 9:00 P.M., 12 hours after her accident, she suffered acute respiratory distress. She died at 1:00 A.M. A white frothy fluid was issuing from her mouth and nose at the time of death. The scorpion that killed her was not the more dangerous of India's two seriously toxic genera. *Buthus* may contain species five times as toxic as *Palamneus* sp.

Early in 1973 my wife and I were at a campsite in southern Angola, midway between the Moçamedes and the Namibe deserts. (The Namibe is an extension of the Kalahari.) We encountered scorpions nightly. They were large and often aggressive.

On the day of our arrival at the site it rained and since it was the first rain in two years we impressed the local citizenry. It was perhaps because of that they appeared beside our dinner table a few evenings later with a teen-age boy who had had the misfortune to sit on a scorpion. While the entire village looked on and discussed the matter I treated the young man's very sore rear end and gave him a good healthy dose of antihistiminics. He survived although his leg appeared stiff the following morning. He was able to walk but with some apparent difficulty.

The fact that we had brought the rain and treated one of their tribesman's scorpion sting really impressed the locals and they staged a dance in our behalf. Five bottles of wine were contributed to help them achieve the preferred state of euphoria. I

The stinger of *Hadrurus arizonensis*, a North American scorpion. The stinger has two openings leading to two venom glands clamped to the hard outer wall of the telson by muscle bundles.

noticed that the scorpion victim had recovered enough in just over 24 hours to participate in a very athletic display. I subsequently lost the specimens of scorpion that I collected and never did determine what species they represented.

Envenomation by scorpions entails a powerful nerve toxin and the local action of a substance known as 5-hydroxytryptamine. It is simpler, as we shall be seeing, than the chemistry and pathological physiology of reptile venoms. The neurotoxin seems to work selectively on the sympathetic and parasympathetic autonomic centers in the hypothalmus. This could explain the sense of profound anguish the victim develops as his symptoms progress. It was once thought that the psychological stress exhibited by victims of scorpion sting was caused by their knowledge of what could follow. Although, naturally the victim of any serious chemical trauma may feel dread and fear, the changes witnessed in scorpion-sting cases are caused by the same chemical that is causing the other symptoms. In this case anguish is not a *result* of the sting symptoms, but is one of them.

Fortunately, a number of antivenins for scorpion stings are available. Hebrew University in Jerusalem produces an anti-scorpion serum, and the Poisonous Animals Research Laboratory in Tempe, Arizona, produces serum covering species of the American Southwest.* Serums are also produced in Cairo, Ankara, Algiers, Mexico City, and Sao Paulo, Brazil.

The concentration of scorpions in an area can be very dense, leading to an unfortunate frequency of painful or even lethal nocturnal encounters. Baerg (1961) reports an episode from Mexico: The city of Durango is inhabited by several species of dangerous scorpions. Between 1895 and 1926 there were 1,608 scorpion-sting deaths in the one city. The majority were children from the age of one to seven. When a bounty for scorpions was offered the boys of the town spread out to seek their fortune. In a three-month period, 100,000 specimens were turned in. The harvest was repeated again and again. (Scorpion deaths for all of Mexico, from 1940 to 1949, numbered 17,750.)

Virtually all scorpion stingings that prove fatal are caused by species representing eight genera: *Centruroides, Tityus, Androctonus, Buthacus, Leirus, Buthotus, Buthus,* and *Parabuthus.* As we have seen from Deoras' report, *Palamneus* can also kill although a lethal case is apparently rare with this genus.

In the United States, *Centruroides sculpturatus* and *C. gertschi* can both kill while scorpions of the genera *Diplocentrus, Hadrurus, Vejovis* and *Superstitionia* are all to some degree troublesome or dangerous. Mexico's horrific scorpion-sting mortality figures are

---

*Two species of *Centruroides—sculpturatus* and *gertschi*—are covered by one goat serum preparation. Two others are offered in an emergency—*Centruroides gracilis* and *Hadrurus arizonensis*—both from rabbit serum.

the work of six forms: *Centruroides noxius, C. l. limpidus, C. l. tecomanus, C. elegans, C. i. ingamatus,* and *C. s. suffusus.* In South America, a broad area to consider as a unit, at least four members of the genus *Tityus* have been implicated in human fatalities: *T. bahiensis, serrulatus, trinitatis,* and *trivittatus.* In India, *Buthus* and *Palamneus* both contain apparently lethal species while *Heterometrus* and *Androctonus* must be considered dangerous. Africa and the Middle East have a number of deadly species belonging to the genera *Buthus, Opistophthalmus, Hadogenes, Parabuthus, Buthotus, Androctonus, Buthacus,* and *Leirus. Scorpio, Euscorpius,* and *Pandinus* range from troublesome to seriously venomous. In addition to the preceding there are at least 20 other genera with an indeterminate number of species and forms that are capable of causing local reactions of varying intensity.

Most of the 650 species of scorpions are not deadly to man, yet six forms could kill 17,750 Mexicans in a decade. A number of factors contribute to the fear people have of scorpions besides the undeniable fact of their toxic nature. Their size is a factor. They are so small (and so secretive thereby) that they can be almost impossible to avoid under certain circumstances. Many forms do regularly invade homes and encounters are inadvertent, presumably on the part of man and animal alike. They do crawl into boots and shoes, and they do get into bed linen.

The fact that scorpions are nocturnal increases the measure of their threat in fact and in the imaginations of people who live in scorpion country. But, perhaps the most potent factor in the scorpion fear-syndrome is the threat they represent to children. A large percentage of the stingings do involve children and children, of course, are generally more sensitive than adults. Suffering children create an aura of dread around an offending organism. In one series of studies done in India, it was estimated that fatalities among children stung by a lethal species of scorpion (*Buthus* sp.) may be as high as 60 percent.

It is unlikely that the time will come when man will not fear scorpions. We will probably always kill all the scorpions we encounter. Yet, most are helpful rather than harmful. They live on insects. Insects are the more harmful of the two.

OTHER VENOMOUS ARACHNIDS

Before going on to other groups, a few additional Arachnids and related animals are worthy of passing mention. Some are truly venomous, others are only so in legend. In the latter category are the solifugae (class Arachnida, order Solifugae), sometimes called "false spiders." They are also called venomous. There is no more direct approach to the subject than to say they are not. The age-old stories are just that, age-old stories, and like almost all tales of that description they are safely disregarded.

The arachnid order Thelyphonida contains the whiptail or

whip scorpions. Also known as vinegaroons from the acetic acid smell present when they are crushed, these animals are persecuted as seriously venomous and dangerous to man. They have no venom glands but do secrete an acidic repugnatorial substance when disturbed (as do skunks.) They are quite harmless, interesting little animals.

The arachnid order Pseudoscorpiones contains the carnivorous false scorpions. Unlike the true scorpions these small animals have no caudal sting, but they do have venom glands. Their pedipalps or pincers are highly developed and look like the claws of a scorpion. They are powerful and prehensile. The immovable finger is lined with a row of cutting teeth and the last of these is enlarged. Through this last "tooth" passes a duct leading to an elongated venom gland contained within the finger itself. In some families both fingers contain venom glands.

The false scorpions are very retiring and have few enemies. They do not hunt but apparently wait for prey, which they can identify by the sense of touch, to come to them. It is thought that their venom plays a role in food-getting and it is to be assumed that it is also defensive. Very little is known about these animals and even less about their venom. The venom is probably harmless to man or very mildly toxic, but that must remain an assumption. Certainly there are few natural encounters where this theory might be tested.

Certain ticks (true arachnids) secrete a neurotoxic substance in their saliva while feeding, and this substance can produce a blockage at the neuromuscular junction and the synapses of the spinal cord. It is apparently produced by the female tick only and can result in an ascending flaccid paralysis. Symptoms, which may not appear until five or six days after the bite, lead to total immobility. Tick paralysis can lead to respiratory failure and death; the case fatality rate is about 12 percent. Man seems to be particularly sensitive to this neurotoxin.

Some mites, too, produce a substance at least annoying to man. The grain itch mite, *Pyemotes ventricosus*, is a minute organism but does bite and cause an urticarial lesion in man. There can be severe infestations and the victim can suffer generalized symptoms, including fever, severe headache, nausea, vomiting, joint pains, and an asthmatic-like syndrome. Loss of appetite and general weakness may also result. The substance causing these symptoms and the reason for the animal to produce it are not known.

CENTIPEDES AND MILLIPEDES

We will acknowledge two other groups of animals in this chapter although they are *not* arachnids. Our reason for doing so is simply that they do not warrant a chapter of their own. As arthropods it is academic whether they appear in this chapter or in the one on insect venoms.

The centipedes belong to the class Chilopoda. They are often large and generally carnivorous animals. They can be as much as 13 inches long, although most species are smaller than that. They are shiny, highly active, and frequently handsomely colored. They are also venomous. The centipedes have one pair of legs on each body segment (21 to 23 in number) and this distinguishes them from the millipedes we will refer to shortly.

The first body segment of the centipede has a pair of modified appendages known as a *telopodites*. Venom glands are contained in the basal portions of these telopodites and ducts run out to sharp and horny tips. The two inward pointing injectors work horizontally and clamp and intrude upon prey while the venom is introduced. Powerful muscles operate the pinching-jabbing mechanism while adductor muscles constrict the venom from the central lumen of the gland into the canal. In a representative of the larger genera the venom gland may be .295 x .039 inches. The duct, generally, will be about as long as the gland. The gland consists of four layers, including the central lumen, secreting cells, a basement membrane, and a muscular layer for constriction. It is a complex structure.

Little is known about the venom of the centipedes and much of what is on record is contradictory. Fable and fact have become mixed and a great deal of work will be necessary to define and separate the two. Intense local pain is usually recorded when a bite occurs and recovery from the resulting ulceration can be slow. There can be lymphangitis, edema, inflammation, and a local necrotic condition leading to the ulceration. Generalized symptoms reported from some centipede bites have included anxiety, vomiting, irregular pulse, dizziness, and severe headache. The anxiety is understandable. Centipedes are not usually pleasant animals to find on one's skin. There have been six to eight reports of the genus *Scolopendra* claiming human lives. I heard such reports myself while in India although I could find no verification of them.

There seems little point in attempting a finite statement or point of view. The *Scolopendra*, at least, or perhaps only a few species of that genus, may be seriously venomous. They may be capable of killing animals as large as man. For the centipedes as a whole, however, a label of mildly venomous would seem to suffice.

A large tropical centipede *Scolopendra subspiripes* from the Bahamas. Note the one pair of legs per body segment. Some species are apparently seriously venomous although the actual degree of toxicity warrants further delineation.

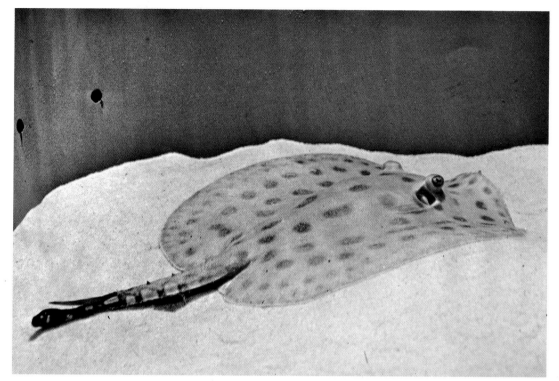

The freshwater stingray *Paratrygon laticeps* is one of the species that has helped make the stingrays the most troublesome of all groups of fish. More people are injured by stingrays than by all other fish combined.

The lion or turkey fish *(Pterois volans)* is a colorful relative of the excessively drab stonefish. It is a seriously venomous species and is now popular with home aquarists. It is not to be underestimated.

The stonefish *Synanceja,* the most seriously venomous of all known fish produces what may be the most agonizingly painful of all known venoms. This animal can be all but impossible to see among the bottom rubble and can dry out at low tide and survive.

The Portuguese man-of-war, *Physalia physalis,* a colonial hydroid that can be very toxic to man on contact. Seen here consuming a fish it has stung to death with its stinging cells.

The millipedes belong to the class Diplopoda. They have two pairs of legs on each body segment. They are vegetarians and do not bite despite the thousand tales to the contrary. Like the centipedes the millipedes are unpleasant for most people to contemplate. It is apparently assumed, therefore, that they must be dangerous.

Although they do not bite the millipedes are equipped with repugnatorial, sac-shaped glands which discharge into a lumen that leads in turn to a foramen or opening. The glands themselves cannot be compressed, but the openings into the lumen are regulated by special muscles. When disturbed the millipedes exude their obnoxious fluid. Some species are able to jet the fluid apparently by fine control over the muscles surrounding the canal or duct. The Haitian species *Rhinocricus lethifer* is said to be able to send its fluid a distance of 33 inches. The fluid burns when it touches the eye and can cause a blackening followed by a blistering of the skin. It creates a slow-healing wound. Although they are certainly not dangerously venomous in the true sense of the word, the millipedes do produce a disagreeable substance which they are able to utilize in a most unique way. The substance, which contains phenols and hydrocyanic acid, is not unlike materials found in every chemical laboratory. That cannot be said of true venoms.

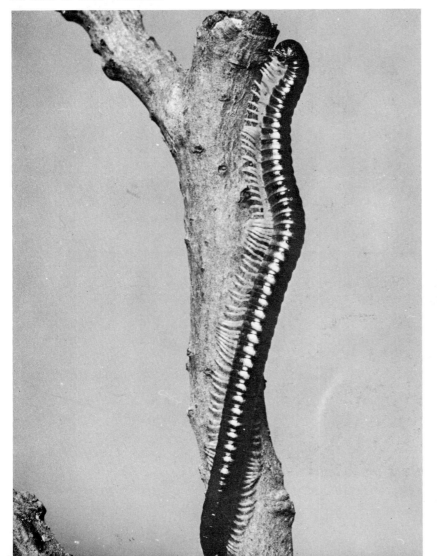

The millipede *Tachypodoiulus albipes*, a perfectly harmless animal much maligned by legend and superstition. Note the two pairs of legs per body segment that immediately distinguish it from a centipede.

Chapter 5

THE

INSECTS

hatever the angle of approach, when we
come to the subject of insects the mind
boggles. No one knows how many kinds
of insects there are, but the number is
certainly more than a million, perhaps
very much in excess of that figure.* It is
quite possible that the full number may never be known. Insects
are found almost everywhere that life exists. A degenerate, wing-
less, mosquito is the largest land animal on the 6,000,000 square
mile Antarctic continent. Insects are found in the burning desert, in cities and swamps, on mountaintops, and in and on other
organisms as parasites.

The insects include the most dangerous animals in the world.
It is certainly true that the flies (and in the broadest sense that
includes the mosquitos) have killed more people indirectly, and
more wildlife, than all venomous animals combined. The victims
of venomous animals of all descriptions that will die this year will
not be a fraction of those that will be killed by fly-and-mosquito-
borne diseases this month. These insects are vectors, carriers,
and intermediary hosts of smaller organisms and they distribute
some of the most serious diseases known to man as they travel
from offal to food supplies, and as the blood-suckers among
them puncture our skins. (The latter form of transmission is the
more serious.) Malaria, salmonella, yellow fever, cholera, sleeping sickness, typhoid and paratyphoid, encephalitis, dengue,

*One entomologist has suggested a figure of ten million!

The paper wasp *Polistes,* an active
and at times aggressive species capable of delivering a painful sting. Individual human beings can exhibit acute
allergic reactions to such an encounter.

many dysenteries, elephantiasis (filariasis), tularemia, leishmaniasis, loaiasis, conjunctivitis, these are just a few of the diseases transmitted by flies and mosquitos. They are found at various stages of their lives in higher animals, including man, in the alimentary tract and even in the urinary tract. As these animals move from piles of filth to stores of edible food they carry endless pathogenic organisms with them. In one study it was noted that a single housefly may deposit 179 pieces of feces and 45 units of vomit—a total of 224 opportunities for disease transmission—within hours after eating from a pile of offal. These figures do not reflect the number of bacteria that may cling to the fly's body, legs, and mouthparts as it moves through our lives. (Again, though, the biting and blood-sucking forms are the most serious carriers of dangerous microbes.) There never has been nor will there ever be venomous animals of any description capable of causing man as much pain and suffering as the flies and their kin, the mosquitos.

The flies, of course, although annoying as blood-sucking animals in many cases, do not inject venom. That capability is left to other insects—the subjects of this chapter—the ants, wasps, hornets, and bees, and certain caterpillars and moths. In many areas of the world the truly venomous insects are more of a health hazard than the far more dramatic and more deeply dreaded venomous snakes.

The class Insecta is divided into 28 orders and of these the Hymenoptera—the ants, bees, wasps, hornets, and relatives—are the most advanced, and also the most diverse. They are not, however, the most numerous in terms of species. Although 100,000 species of Hymenoptera have already been identified (many await identification and description), that number pales before some of the others. The Coleoptera, the beetles, already number 275,000 species and an enormous amount of work remains to be done on the order. The butterflies and moths, the Lepidoptera, already number 200,000 and the surface has just been scratched. Of the really massive orders only the deadly order Diptera, the true flies, trail the bees, wasps, and ants with about 85,000 species so far named.

The word Hymenoptera is derived from two roots—*hymen* or membrane and *pteron* or wing. Typically, members of this order have two pairs of wings that are membranous and transparent, or nearly so. They are also typically venomous.

VENOMOUS ANTS

The ants belong to the family Formicidae in the order Hymenoptera. (The relationship of the family name and *formic acid* will be immediately apparent.) How many species there are is not really known. For years it was usual to suggest 5,000 species, but that figure was rounded off for conversational purposes and does not suggest, much less reflect, the amount of work remaining to be done in our effort to understand these fascinating animals.

It is difficult to speak of ants (or most of the Hymenoptera, for that matter) without being sidetracked by their most interesting single characteristic, their sociability. Only the termites, much more primitive insects of the order Isoptera, have more complex societies than the Hymenoptera. Ants, particularly, construct gargantuan nests, diversify morphologically according to their job assignment, and live their lives as if on some enormous feudal estate. Their social structure is rigid, functioning in some kind of fearful, Orwellian perfection which overrides all considerations of the individual. An ant colony has been likened to a living organism and the individual ants to something like corpuscles in a bloodstream. However insignificant the single ant may be it is often venomous and capable of inflicting a painful sting in defense of self, but more likely in defense of the society in which it dwells.

Not all ants are venomous, of course, but in species where the envenomating apparatus is present it follows the normal hymenopteran pattern and occurs only in the *adult* but sexually undeveloped female. The presence of venom glands in only one sex is not unique to the Hymenoptera, but it is highly unusual. Writers that have referred to it as unique have overlooked the duckbilled platypus, which we will discuss in a later chapter; only the male platypus is venomous. Among the ants, as among the bees, wasps, and hornets, only the female workers, or soldiers, can envenomate.

It is difficult to determine if venom among the ants is primarily defensive—for protection of the nest—and secondarily offensive—for killing insects used as food—or vice versa. Because ants are capable of overwhelming almost any living creature by their staggering numbers, and since everything else about ant society and ant evolution favors the group rather than the individual, I tend to favor the idea that venom among these animals was evolved in the first place as a means of defending the nest. That, however, is an opinion, an assumption perhaps, and not a statement of fact. (Since even the nonpredatory species that utilize only carbohydrates of plant origin in their adult phases feed their larvae on animal food, not all students agree with my conclusion. They hold that the origin of the stinging capability among the Hymenoptera lies in food-getting.)

The venom gland and attendant structures in the Hymenoptera evolved from the female reproductive system, the ovipositor through which the female deposits her eggs. In groups that do not sting, the ovipositor looks so much like a stinging barb that for a long time their innocent nature was not appreciated.

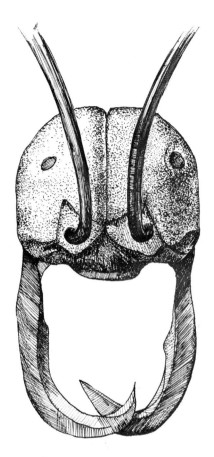

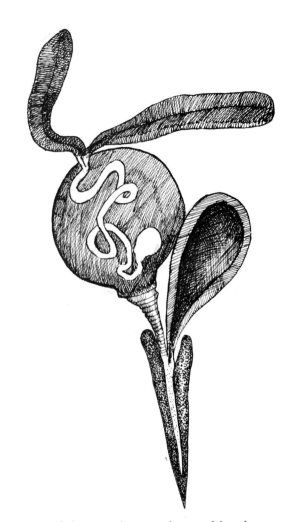

Simplified ant structures for comparison. At left is the biting apparatus of an idealized non-venomous ant, while at right is the equally idealized venom apparatus from a species equipped with this powerful chemical weapon.

The envenomating system of the ants is complex and has been subjected to intense study. In its fully developed form, the stinging system consists of one or more venom glands, a reservoir for the storage of venom, and a device known as Dufour's gland. This gland is closely associated with the act of envenomation yet the part it plays is not really understood. Some investigators say that it is a lubricating device to make it easier for the ant to insert its barb. It is frequently called the "alkaline" or "accessory" gland. The reservoir may be large and extend forward in the ant's abdominal section. Ducts from the reservoir and from Dufour's gland pass back to the tip end of the segment, or *gaster*, and supply a chitinous sting. In some species a pair of "sting" glands are attached to the base of the barb.

There are so many forms of ants, at such widely divergent levels of evolution and degeneration, meaningful generalizations are extremely difficult to find. The venom reservoirs in some species are sac-shaped and asymmetrical, they are spindle-shaped in others, while in the higher ant forms they seem to be at least roughly spheroid. The thickness of the muscle walls surrounding the reservoir varies to a considerable degree, but the structure seems to be well insulated from the rest of the ant's body. This may be to protect the ant from its own toxin.

The actual sting itself is an interesting if inconstant device. In wood-boring species the ovipositor is saw-like and very tough. In some wood-borers it is used not only as a bore, but also as a probe and an egg-laying tube. In other ants (*Parasitica*, for example) it is used for placing eggs and as a sting for introducing venom. The variations among the ants seem almost endless as one reviews these animals at their various levels of development.

No one seems to know the purpose of the extra sting glands found in some species. In bees, particularly the social bees, it is possible that these glands provide a strong scent or attractant. Presumably their purpose is to mark the area stung with a powerful smell that will draw other bees from the same hive and encourage them to sting, too. However, sting glands are found in both solitary bees and social ants as well. Their real function is not understood.

Any study or reference to ants must take into account their extremely social nature. They signal to each other in a number of ways and chemicals are among the most important of these. Just as an ovipositor may be a wood-boring instrument and an egg-positioning probe, a probe and a stinger, or a stinger and an egg tube, so chemicals may have two or more functions. Any of the substances secreted by the ant in association with envenomation, including the venom itself, may also serve as a chemical signal.

Just to put the ant into a frame of mathematical reference, we are speaking of an animal that may not weigh more than one .000165345th part of a pound—which means that a 200-pound man would weigh as much as 1,209,509 of these creatures. When a complex animal—and an ant is very complex—is that small it is extremely adept at keeping secrets. We have a great deal to learn. (On the subject of weight, which is presented above in a kind of mathematical analogy to stress size, most ants weigh between 75 and 347 milligrams. We chose the bottom weight for our analogy and used the standard formula of mg x 2.2046 x 10^{-6} = pounds. This measure somehow seems more effective than saying an ant is small, or a third of an inch long. Weight, since it does not involve an orientation to height, length, or width, seems to project *size* in a way that can be universally appreciated.)

In a recent study (Wilson and Regnier, 1971) it was suggested that Dufour's gland is an integral part of the social ant's alarm-defense system. Its secretions are seen as an ant's means of spreading and enhancing the penetration of the formic acid coming down from the venom reservoir and, when combined with the mandibular gland in the ant's head, as an element in the chemical alarm-defense system of the colony. Such a double use, as we have noted, would not be surprising. It may have a third use as well. Certain ants take slaves, literally that. Very aggressive ants out on a slave raid against the nest of another species may discharge decyl, dodecyl, and tetradecyl acetates into the nest at the beginning of their raid. These substances might serve as a kind of panic propaganda and disperse the defenders of the target nest. There is no doubt that just as the evolution of the envenomating apparatus of the Hymenoptera traces back to the female reproductive system, the apparatus and its chemicals as they exist today are strongly linked to signaling and social activities.

The venoms of ants, complex substances that vary greatly from group to group and even from species to species, are seldom fatal to man. They are an annoyance certainly, but seldom seriously harmful. People have died after being stung by ants, but such dire results were due to an allergic reaction and not to the toxicity of ant venom. The *Journal of the American Medical Association* (Sept. 30, 1961) observed in an editorial: "Almost all deaths from hymenopterous venoms are due to anaphylactic shock. Persons highly sensitive to these venoms should be desensitized with emulsified extracts of bee or wasp venom, or a mixture of both, or extracts of the animal's body."

In a review of insects of medical consequence in Texas, Micks (1960) discussed a survey of 75 ant-sting cases treated by 25 physicians. Twelve percent of the victims experienced systemic responses to the ant stings. Five suffered anaphylactic shock (which *can* be swiftly fatal), three had asthmatic rales, swollen faces, lips, and glottis, and massive rashes over their bodies. Fainting, vomiting, and abdominal cramps were seen in a number of the cases. One patient had to be supported with oxygen. Severe pain and local infection were recorded in some of the cases. Obviously, these were cases in which the victims were affected by their own acute sensitivity in combination with the ant venom.

As we will note when we come to the bees, wasps, and hornets, there is the strong suggestion that many more people die from hymenopterous insect stingings than are recorded. There is the suspicion that a number of otherwise unaccountable deaths in the past might be blamed on the acute sensitivity of some people. This is probably truer for the flying insects than for the ants, but we do know that between 1950 and 1959 at least four Americans were killed by ants. Although there is no certain way of predicting outcome, it seems to be generally true that once a person exhibits a sensitivity each subsequent episode will elicit a more severe reaction. The reaction can work the other way in some cases, or even seesaw, but generally it gets worse with each exposure. It is possible that some of those ant-sting victims which show the worst reactions are in fact reacting to what has been a whole long series of exposures. Initial exposures may have caused such mild reactions that the person may have forgotten or, indeed, may never have really known what it was, if anything recognizable, that "bit" him.

Ant venoms, of course, are complex; all venoms are. Depending on species they can contain various proteins, hydroxy acids, histamine, hyaluronidase (spreading factor), amines, formic acid, and a whole series of organic materials of complex chemical structure. Once again one is awed by complexity not only of the morphology, but also of the chemistry of such very, very small creatures.

Perhaps 50 percent of the ant species show a degeneration of the stinging mechanism and it may be on the way out. Many explanations of this degeneration have been suggested, but its causes remain unknown. For ants, apparently, the sting at the rear end of the abdomen is impractical, particularly when the target is a small, fast-moving insect. Ants may be evolving away from their cumbersome stinging system which is complex, fragile and, as we have said, sometimes difficult to employ. They may be moving toward a system that can spray venoms for some inches and incapacitate a foe at least long enough so that it can be grabbed by their mandibles. In some ants Dufour's gland is very much larger than the venom glands and perhaps the mandibular gland in the head is becoming involved. Since ants use a number of chemicals not derived from their venom gland in defense-alarm signaling, it is quite possible these same extra-venom chemicals are figuring more and more as repugnatorial substances and as temporary incapacitators. In this case, at least, the study of venoms is a study in evolution.

BEES, WASPS, AND HORNETS

The bees, wasps, and hornets are better known as stinging insects than the ants. (A great many people, it seems, do not realize that ants sting. They think that ants, with their powerful mandibles, pinch and bite [which many do], but have not generally associated the rear end of the ant's body with pain and annoyance.)

To make a few worthwhile distinctions: Wasps are generally believed to be older than bees and are usually carnivorous. The bees, more highly advanced than the wasps, usually live on plant carbohydrates even in their immature stages. Wasps use their stings defensively, of course, and to subdue prey. Very often animals stung by wasps are not killed, but put into a kind of suspended animation. Their nervous systems are destroyed by the venom and the wasps lay their eggs on the paralyzed carcass. Their larvae, upon hatching, thus find a store of food that will see them through to the next stage of development.

Bees that live on plant substances as mature animals and feed their young on honey and pollen also may have a stinging apparatus. This apparatus is not vestigial in bees, not a hangover from earlier levels of evolution when they were more closely related to predatory wasps, but a defensive device for the protection of the hive. Among the social bees and wasps, as among the ants, the nest or hive is all. Individual animals are engineered to serve the colony and females of certain castes are equipped to serve as defenders. Not at all surprisingly, wasps that use their stingers for food-getting retract their lancet after envenomation and live to use it again. Bees that only require their venom in the case of warfare cannot recover their stinger from heavy-bodied enemies and die shortly after bringing it into play.* The evolutionary or survival advantage to a hive of killing off those defenders who prove themselves willing and able to do battle with a large enemy is not clear. It is just as if an army took all those men who actually participated in a battle after the fight was over and killed them, leaving an army of only those men who would not, could not, or otherwise did not actually kill or attempt to kill the enemy. In the long run our society might be better off if we did not have the ability to kill locked into our gene pool, but how or why this works for bees is not clear. It would seem, on the surface of things, to be counterproductive.

Since bees that die after they sting are descended from animals that did not die after stinging (as far as we know), we may assume that an advantage did or does exist for this strange change to evolve. (It is impossible for a carnivorous species to have this feature because, of course, they would die after obtaining each meal.) We must remember that the workers that defend the hive against attack do not figure into the genetics of the species—they are sexually undeveloped females. The females that do reproduce, the queens, would not be called upon to fight off an enemy. (Queens do have stingers and in some species an emerging queen assumes her role only after killing all other young queens.) Could it be that a worker or defender, if you will, once she has used her stinger to kill even in defense of the colony is then too aggressive and could be dangerous to have around? This idea is posed as a possible explanation, not a statement of fact.

Bee venom is made up of a complex mixture of polypeptides, with proteins as spreading factors (hyaluronidase), enzymes like phospholipase, and histamine. There are differences of both quantity and kind of venom among the diverse groups of bees.

The venoms of the Hymenoptera were evolved for use against other insects. Originally, it is assumed, they were used princi-

---

*They can apparently survive if the animal they sting is small and "thin-skinned." In experiments where bees were induced to sting through a thin membrane, mortality was only 10 percent.

pally in predation, to obtain prey by killing it or at least paralyzing it. In the solitary species the main use may be largely defensive, to avoid or counteract molestation. In the social species defense of the colony is the issue. The role the Dufour gland plays in all of this is as mysterious as it is among the ants. The discharge from the Dufour gland is volatile and contains hydrocarbons, ketones, and esters. They apparently act as spreading agents, helping the substance from the venom glands penetrate and reach maximum effectiveness quickly. The Dufour gland content may be a venom in its own right, or this may be true only in some species.

Benton and his associates (1964) showed that mice fed on a high protein diet just before or for three days before being injected with honeybee (*Apis mellifera*) venom were more susceptible to the venom by a factor of at least two to one. Mice on a protein starvation diet for three days, conversely, had a much better chance of surviving venom injections. The doors opened by these observations lead to a veritable maze of ideas and possibilities.

In the past studies of bee venom were carried out with a substance made of entire bees ground up, or just the abdomen pulverized and homogenized. Now it is possible to extract pure venom using sophisticated electrical techniques, and it has been found that the material obtained is quite different from the older, much less refined substance. Honeybee venom is fairly constant in toxicity from one part of the country to another, but appears to be more powerful during the spring and summer than during the other seasons. It has the power to elicit dire allergic reactions in some higher animals.

Simplified, dissected view of the beesting apparatus. In the plant-food-eating bees this defensive device is usually torn out of the abdomen when it is used and left in the animal attacked. The bee dies as a result of the trauma.

Multiple stings by colonial bees are not uncommon and can result from the most inadvertent encounter. It is known that swarms of the giant Indian bee (*Apis dorsata*) have killed water buffalo and even elephants. Parrish (1963) did an analysis of the 460 deaths that occurred in the United States between 1950 and 1959 as a result of contact with venomous animals and listed 225 for the bees, wasps, and hornets. In that same period of time snakes killed 138. Here is the list for the decade of the Fifties adopted from Parrish:

VENOM DEATHS 1950-59—UNITED STATES

Bees	124
Wasps	69
Yellow Jackets	22
Hornets	10
Ants	4
Snakes	138
Spiders	65
Scorpions	8
Others	20
	460

Two hundred and twenty-nine people killed by Hymenopterous insects in a decade is not inconsequential. It is not epidemical either.

Of the 229 Hymenopterous sting deaths, eight resulted from hundreds of stings in the same individual. A person overwhelmed by hundreds of bees or wasps need not depend on an allergic reaction for death to overtake him. He will be killed by the venom itself. In Ceylon, I saw cages set along a path for people to climb into in case of swarming bees. There the jungle bees (*Apis* sp.) are so aggressive and their venom so painful and potentially dangerous that provisions must be made for sheltering people in the event of an attack. At the giant rock fortress of Sigiriya, where thousands of tourists go each year, there are cages marked "In Case of Bee Attack" in three languages. The same is true at Adam's Peak.

The National Safety Council has long felt that bee and wasp stings probably account for a number of otherwise unexplained deaths each year. Acutely sensitive people driving along at 50 or 60 miles an hour may be stung by a frightened insect which has gotten into their car and be unable to bring the car to a stop in time. Such people, hitting an abutment or going over an embankment in their cars, have probably been listed as heart attack victims in the past.

In his survey of the fifties, Parrish noted: "Hymenoptera insects kill allergic patients with alarming suddenness. Most of these people died suddenly within 15 to 30 minutes, and were seen by physicians only after death. Many of these victims did not have time to summon medical aid, owing to the overpowering effects of laryngeal edema or anaphylactic shock, or both. Physicians certifying these deaths often wrote that the patient died from 'shock,' 'allergy,' 'anaphylaxis,' 'laryngeal edema,' 'heart failure,' 'coronary occlusion,' or 'stroke' following a sting or multiple stings by Hymenoptera insects."

Fluno (1961) suggested a classification for the reactions people get when stung by a bee, hornet, or wasp: *Hymenopterism Vulgaris* for painful local reactions, *Hymenopterism Intermedia* for symptoms spread beyond the area imediately around the sting and when systemic reactions are evident, and *Hymenopterism Ultima* for lethal or near-lethal reactions. He suggests that people exhibiting the first two levels of reaction may be affected by only certain elements of the complex protein. Those suffering from *H. Ultima* may be reacting to substances that do not even touch victims in less acute phases.

We mentioned earlier the alarm-defense chemical signals of ants and how venom and communication were interlocked. The same is true, apparently, of the wasps and bees. At least one rather humorous episode attesting to a signaling system among bees is on record. Morse and Benton (1964) used an apiary at Cornell for venom collection—using their electrical technique that inspires bees to sting through a fine membrane. The apiary was 40 years old at the time of the first experiments, and there was no record of anyone not actually handling the bees having been stung. Apparently, so much of the alarm signal was discharged by the bees being milked that the entire area surrounding the apiary was reduced to a near battle zone. Women using a swimming hole a quarter of a mile away were attacked, as were students using pathways in the area and people in a parking lot some distance away. The alarm or attack signal was apparently very inspiring to bees in the apiary that had not been milked. After this incident, it was clear that the experiments could be carried on only at some distance from the campus. The bees were moved during the night.

When a bee stings its barbed lancet becomes fixed in the tissues of a "thick-skinned" victim. As the bee pulls away, its poison sac, muscles, and other attendant viscera are torn off—resulting, of course, in the death of the bee. The muscles attached to the sting and glands may continue to pump venom down through the sting for as much as 20 minutes. People attempting to remove the sting by grasping it are in danger of forcing even more venom into their flesh by the pressure they exert. Stings visible on the skin should be scraped away with a knife blade. There is no antivenin available for bee sting and, indeed, one would be of

A bee's stinger magnified 250 times. The rear-pointing barbs make it difficult to withdraw the stinger once it has been inserted and it is usually left lodged in the victim's tissues. The damage done to the bee in the loss is fatal in most instances.

A bee's stinger embedded in a man's hand—greatly magnified. The venom glands that have been ripped out of the doomed bee's body can be seen still attached to the stinger. Any attempt to grasp the stinger would result in the glands being squeezed and more venom being forced down into the wound. Such relics should be scraped away with a knife blade.

little use. People suffering mild reactions do not need one. Those suffering "ultima" reactions would not have time to get it.

Hyperallergic people can often be desensitized, and anyone exhibiting unusually severe reactions to insect stings should discuss the matter with a specialist. (It is also advisable to carry emergency medications.) I personally know of people who cannot go near flowers or into the country during the spring, summer, or fall for fear of being stung by a bee. Theirs is a very unpleasant sensitivity to have to live with.

The fact that a man can be killed or made critically ill by something less than a millionth his own weight is bizarre. It is also indicative of the refinement of the ability to envenomate as well as man's ability to become sensitized to foreign proteins. Among bees whose venom is strictly defensive, the power to incapacitate larger animals is perhaps understandable. A bee never knows what may attempt to attack its hive. In the predatory wasps, however, there is some question about all of this. It is assumed that the venom is *primarily* for food-getting, if not for the adult then for an immature wasp whose egg will hatch on the paralyzed body of its first food supply. Surely a venom that can incapacitate a man is more than a wasp needs to destroy the nervous sytem of a caterpillar. Of course wasps, too, defend their colony, and it is possible that although wasp venom is a food-getting device in the first place its strength is linked more to its secondary use, defense. We again deal with assumptions.

As everyone knows *pain* is usually the first indication of a bee or wasp sting. The origin of that pain is interesting and open to some question. It is quite possible that the substances injected by wasps and bees do not cause pain directly but rather create the conditions for pain. For example, *mast* cells are widely distributed in connective tissues and they contain histamine. If a cell is injured by bee venom and histamine is released, itching can result. At certain concentrations the itching can turn to pain. A very powerful pain-producing substance is 5-hydroxytryptamine (5-HT)* which is found in blood platelets. If these cells disintegrate (and this can happen when they come into contact with injured tissue), the stored up 5-HT could be released in sufficient quantity to cause pain. Potassium is more heavily concentrated inside of cells than it is in the fluid surrounding them. The destruction of a cell wall could release potassium in large enough amounts to effect nearby nerve endings, causing acute pain and distress. Much more work must be done in this field, but it may be at least partly true to say that bees and wasps inject a venom that makes people hurt themselves with their own available chemistry.

*Present in large amounts in some hymenopteron venoms, e.g., hornets.

URTICATING CATERPILLARS

The caterpillars of several different families of Lepidoptera (moths and butterflies) are able to envenomate simply by brushing against other animals. Their venom system is defensive, used to keep predators away. The envenomation takes place through hairs and spines of several different kinds, devices designed just for this purpose as far as we know.

Normal insects' hairs are relatively soft and have fairly blunt tips. They are not brittle and do not puncture skin and break off in it, and they are not barbed. Still, some people are allergic to these normal hairs and break out in a distressing urticarial rash whenever they come in contact with them. The mechanism involved is not understood at all.

Nettling hairs of some families look like *normal hairs* until seen under high magnification. They are then seen to have overlapping plates with the free end pointed backwards toward the thickened base end. These hairs are brittle enough to break off in a person's skin and can prove to be an irritant.

Spiegelhaare are short thin hairs with pointed bases. These spicule-like devices pull out when the caterpillar brushes against a victim and stick into its flesh. There are at least three different kinds of *spiegelhaare* recognized and all are highly irritating. In some forms these small hairs are held in bunches and released like a shower of little darts when the animal is bumped. It has been estimated that 750,000 of them may be released in a single encounter. The mechanism by which this is accomplished is not known nor is it known if the hairs are venomous. They are, though, highly irritating, particularly if they get into the eyes, nose, and throat.

Brush hairs appear on the backs and sides of some species and are longer than the *spiegelhaare*. They are usually in bunches and look distinctly brush-like. One bunch may have several thousand hairs in it. The hairs are barbed and sharp at each end. They are said to feel like the sting of a blood-sucking insect when they jab into the flesh and hook there.

(Imaginal Anal Tuft Hairs are found in adult moths of both sexes, but appear to be more highly developed in the females. They are urticating and painful.)

In addition to the urticating hairs, some forms of caterpillars have poisonous spines. These are quite different from the much simpler irritating hairs. They are not derived from a single enlarged hypodermal cell the way a hair is, but are an extension, of

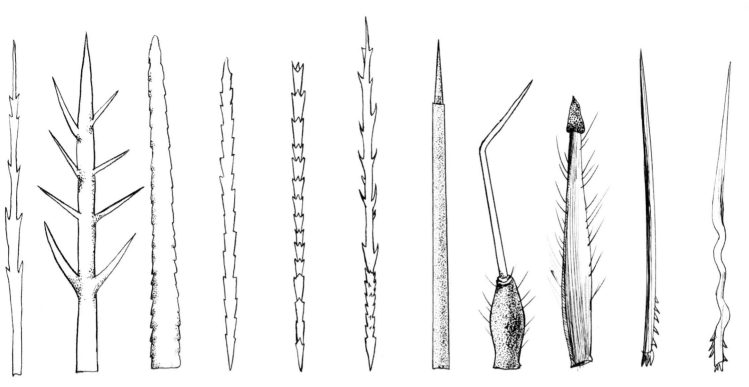

sorts, of the body wall. Moreover, they are not distributed generally over the animal the way hairs often are, but are bunched on tubercles on the sides and backs. Each spine is hollow and at its very tip has a fine opening leading from the central cavity to the outside. Where the spine meets the caterpillar's body there is an expanded bulb-like cavity filled with venom. When the animals are bumped or otherwise attacked the fluid is forced out of the bulb into the canal and then into the victim as the spine breaks away, imbedded in the enemy's flesh. The mechanism is so delicate that enough pressure is exerted by a simple brush to cause the venom to be expelled. In addition to the pressure of contact many caterpillars are able to exert extra internal pressure against the bulb, ejecting even more venom. Some species have erectile muscles exercising some measure of control over the spines. In some species there are sensory hairs or spines that extend beyond the venomous spines ready to alert the mechanism that it is about to be bumped or attacked. Some forms of sensory hairs are called whips, but they are too fine and soft to penetrate tissue.

In addition to having hairs and penetrating spines, some caterpillars exude a liquid through easily broken spines or special pores. This irritating substance can cause welts or rashes.

The venomous apparatus of the Lepidoptera is defensive. It plays no role in food-getting, but rather protects the animal from being used as food. If an individual caterpillar is snatched off a twig by a young and inexperienced bird or insect-eating mammal, it will probably be too late for that caterpillar. While making the would-be caterpillar eater decidedly uncomfortable, it may be suffering mortal injury itself. A bird's beak acts quickly, too

For comparison, design variation among caterpillar hairs and barbs. Although all are drawn to one size here these barbs vary greatly in size and in the clusters into which they fit. These devices come away from the caterpillar upon simple contact and cause varying degrees of discomfort.

quickly perhaps for bird and caterpillar alike. But the bird learns a lesson. Caterpillars are very distinctly marked animals and the bird once it has gotten a mouthful of spines and hairs will be unlikely to try the next caterpillar of that design it encounters. So, although a caterpillar may be crushed to death, its defensive spines and hairs will have been of use to its species.

In a review of insect injuries in Texas, Micks (1960) discussed 54 cases of caterpillar envenomation. Forty-seven reported severe local pain and ensuing dermatitis. Seven patients suffered anaphylaxis, fainting, urticaria over much of the body, nausea, and vomiting. The lymph system reflected the insult in about half the cases. In the United States the puss caterpillar (*Megalopyge opercularis*) is responsible for many clinically-seen cases each year. Systemic reactions as well as severe to agonizing local ones are reported. While not lethal and not of enormous significance, these caterpillars are annoying and even dangerous to some people.

For purposes of clarity, people envenomated by adult moths are said to suffer from *lepidopterism*. Adult forms of some relatively few species are equipped with tufts of venom-carrying hairs or *flechettes*. These devices are, as in the caterpillars, defensive. When it is the immature animal, the caterpillar, that causes the envenomation the disease is called *erucism*. The ability to inflict this disease is found among caterpillars over much of the world. Some of their venoms are so powerful that cattle and pigs eating caterpillars along with fodder have been known to die from internal envenomation. In these cases the venom becomes an economic and agricultural problem.

Cases where hairless and spineless caterpillars are involved, but where an irritating fluid comes into contact with another animal's skin, are known as *para-erucism*. Some caterpillars weave venomous spines into their cocoons as they go into their pupal stage. These spines, of course, protect them when they are otherwise totally helpless. Emerging adult moths can pick up the spines and impart them upon contact with another organism. This inherited ability to envenomate is called *meta-erucism*. The reverse works as well. Female moths may deposit spines with venom on their eggs. The emerging caterpillar can pick them up and carry them, imparting venom to anything molesting it. This is called *pseudo-erucism*, but is really a kind of false or consigned *lepidopterism*.

Caterpillars and moths do not create what may be considered major health problems, yet they constitute an annoyance and only rarely a medical emergency. Keegan reports that on the island of Honshu, in one month during the summer of 1956, 250,000 people were affected by the moth *Euproctis flava*, a known venomous species. In other areas, as a result of infesta-

The puss moth caterpillar *Cerura*. In the larval form this species is capable of causing a distressing urticarial rash by pricking a victim with sharp, venomous bristles.

tion by other species, there have been similar outbreaks of skin disease, discomfort, and some genuinely ill people. This condition falls far short of the emergencies created by spider, scorpion, and snake envenomation, but it is an added burden for a great many people.

A WORLD OF VENOMS

When compared to the arachnids or the snakes, the insects do not seem very dangerous—in a one-to-one comparison. But, insects never seem to come that way—one at a time. Insects exist in numbers that we cannot really fathom. We speak as if we could but we do not, not really. It is beyond us to comprehend a single ant colony with a population equivalent to that of three nations.

Just as the numbers of insects befuddle us so do their social organization and their individual dedication to colony defense. We do not really understand the ant or the bee that will throw itself without a moment's hesitation into a situation from which it cannot possibly emerge. It is programmed to do so, of course; it has no choice in the matter. It is a cell in a larger organism—the colony or hive. The insect world is a kind of nightmare world of billions of venomous animals locked in combat in which uncounted numbers will die. Yet, that world is there. It is larger than ours, more diverse, and stronger; it not only outnumbers us in individuals but also outweighs us, so enormous are its numbers.

Occasionally we come up against the venoms of these motes of life and an animal that we outweigh more than a million times sickens us and has even been known to kill. We brush against a bizarre but somehow beautiful caterpillar and are reminded by our discomfort that the caterpillar, too, was designed to survive. Of the more than a million kinds of insects presently known or projected to be known in the future, only a small number —perhaps not more than a few score thousand—can envenomate us to any degree we can recognize. Those that do remind us that "down there" there is a whole world of venoms that we fortunately need never worry about. Our inability to interact, however, does not deny that other world's existence. It just makes it of less consequence in our own desire to survive.

POSTSCRIPT

During the early phases of the research for this book (it took four years altogether), an article appeared in the *New York Times* about a "bug" that had been discovered in Israel that was believed deadly to man. Other papers picked the story up and the exaggerations began to mount. I wrote to Professor Shulov of Hebrew University in Jerusalem and he was kind enough to send me two specimens of the insect. He also was blisteringly angry at the media for breaking the story before all the substantiating research was done.

After two long delays I managed to fly to Israel and visit with the professor in his office at the University, and at his Biblical Zoo. The strange and still unexplained story of "afrur" began to emerge.

Afrur is the Hebrew word for dirt and may refer to a part of the behavior of the insect *Holotrichius innesi*. This insect (*afrur* is apparently its only known common name) belongs to the family Reduviidae and is thus related to the bedbugs. There is some suggestion that all members of the family are venomous to some degree but certainly not to the extent of this species.

The venom of *H. innesi* is introduced through a movable proboscis and has powerful neurotoxic as well as hemolytic elements. I heard it suggested that the hemolytic constituents would be academic since the neurotoxins would kill the victim (human and otherwise) before damage could be done to blood tissue. It is believed by Dr. Shulov and his associates that this venom is the most potent in the Middle East, which is saying rather a lot. Although there were no certain human deaths attributed to *H. innesi* at the time of my inquiry (fall 1972), when I asked for a comparison with other known venomous animals of the region I was told *Echis* would be "a mosquito by comparison." When we discuss the saw-scaled viper the force of that statement will become apparent.

Experiments with laboratory animals show paralysis and asphyxia to be symptoms of envenomation by this bug. There is cardiac involvement, no conspicuous necrosis, and pain is doubtful. The female is apparently much more venomous than the male; death follows envenomation very rapidly in experimental animals. A mouse that might survive the bite of a cobra for several minutes will die almost instantly when jabbed by one of these insects.

The amount of venom administered is thought to be large for an insect an inch or less in length. There are three venom glands to service the single stinging mechanism in the proboscis, or extended mouthparts. Two small glands are located in the thorax and one enormous one in the abdomen. The female is broad and flat in comparison to the long and slender male. Her broad abdomen no doubt accommodates the extra large venom gland that makes the female so toxic.

There are a number of significant features to the study of *H. innesi*. Its degree of toxicity is stunning. It may lead to the study of other insects whose venomous nature has hardly been suspected, and may also play a possibly unsuspected role in the public health of the Middle Eastern region. This species is believed to be limited to the Sinai and Negev deserts in Israel. It may be found elsewhere, however, and it may have seriously venomous congeners.

The Hebrew name for this insect, *afrur*, may come from its habit of concealment. When exposed (it likes to hide under rocks in the desert), it begins to frantically throw sand onto its back. Special glands on its back produce a glue to hold the sand. In moments it is invisible even to close scrutiny. It is a most interesting animal and is today yielding quantities of venom for study through electrical milking. The principal place of study is the Department of Entomology at Hebrew University. I am grateful to Dr. Shulov and his associates for discussing the subject with me before their research has been completed. The parameters given here must be taken as suggestions only, for all the facts about this newly discovered venomous animal are not in by any means.

Holotrichius innesi, a small bug from
the Sinai desert may be one of the most
seriously venomous creatures on earth.
It's deadly nature was only recently suspected and research is underway at
Hebrew University in Jerusalem. The
dangerous, broad-bodied female is
above.

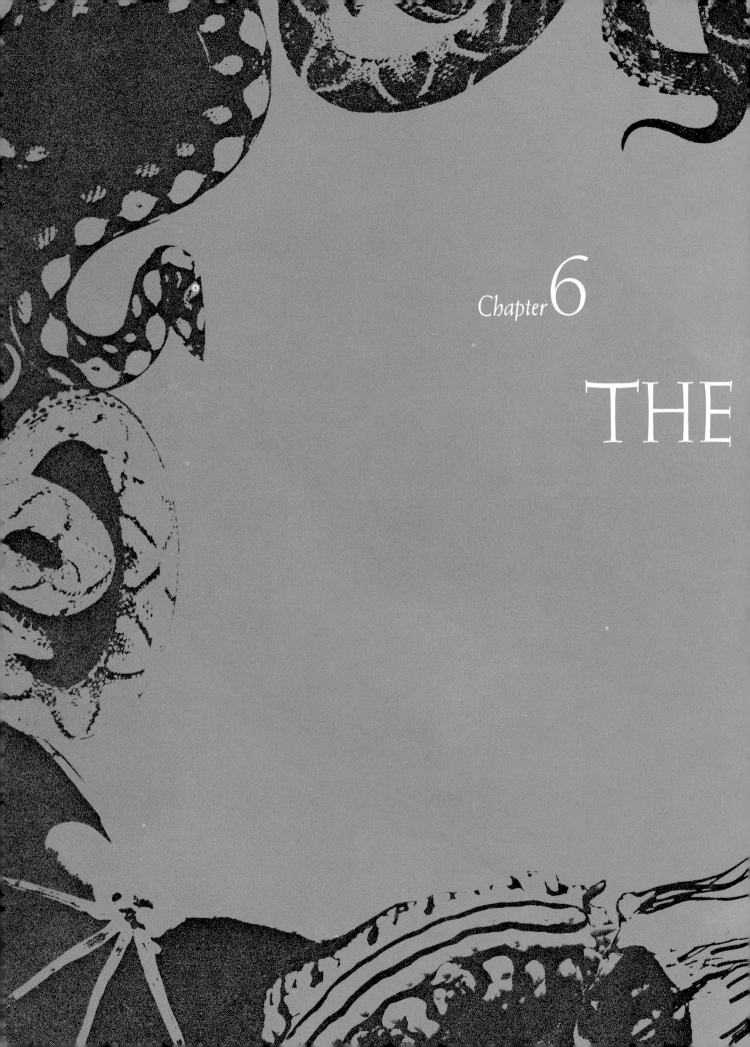

Chapter 6

THE

VENOMOUS
FISH

The scorpion fish *Scorpaena scropa,*
one of a number of species in this
genus capable of causing serious systemic reactions as well as extreme local
pain. A genus very troublesome to man
since they are equipped with a powerful toxin, the sole use of which is self-
defense.

uthorities on marine life estimate that less than 5 percent of the world's venomous fish have been studied except in the most cursory manner. It seems certain that only a fraction of the fish that are venomous have been positively identified as such. In short, we know relatively little about the subject of venomous fishes.

Biotoxic fishes can be grouped into three categories. In one group are the fishes that are poisonous to eat. Although these fishes constitute a major health hazard in many parts of the world, they do not come within the scope of this book. Our discussion will center on a second group of biotoxic fishes, the venomous fishes, animals equipped with a distinct venom apparatus, and also refer to a third, little known group, the ichthyocrinotoxic fishes. Fishes in this latter group lack a venom apparatus, but discharge poisonous substances directly into the water around them.

Dr. Bruce Halstead, in his monumental three-volume work, *Poisonous and Venomous Marine Animals,* provides a classification of the fish known or accused of being venomous. He lists 4 sharks, 58 stingrays, 47 catfish, 4 weeverfish, 57 scorpion fish (including the most venomous fish of all), 15 toadfishes, 3 stargazers, 8 rabbitfish, and 8 surgeonfish. In addition to these he notes that fish in a number of other categories have been accused of being venomous. Some of these *suggested* venomous fish are questionable at best, but they should at least be listed here although no further discussion of them will follow. The list includes deep-sea dragon fish, moray eels, squirrelfish, jacks, butterflyfish, old wives*, mojarras, bugglerfish, sea robins, snappers, finger fish, perch, scats*, sea bass, snake mackerel, sculpins, flying gunards, dragonets, goosefish, and anglerfish. Many of these fish may eventually be identified as venomous, but certainly not all of them.

In all cases that we will be discussing the envenomating apparatus of the fish is part of a defensive system. It is not used even secondarily for food-getting. In many cases, as we will be seeing, envenomation is likely to be the result of a chance encounter. Most species of venomous fish rarely attack. They usually wait to be bumped or grabbed. Some species position themselves to increase the likelihood of contact with their venomous spines when they sense a disturbance nearby. It will be noted, too, that fish do not have a venomous *bite*. They have spines which are associated with venom glands. The idea that venomous spines are less efficient than a snake's fangs will seldom occur to anyone who has stepped on a stringray or kicked a stonefish.

The outcome of virtually all encounters with venomous fish is pain, excruciating unbearable pain that at times has been reputed to drive men mad. This theme is repeated again and again in clinical reports. Most of the venoms, however, are not just local in effect. Some, if not all, are systemic and more than a few can be lethal. With man's exploitation of the sea as a source of food and minerals likely to increase and reach new and more remote areas, contact with venomous fish is also likely to increase. The danger these animals can pose to the unwary is of no small consequence.

Venomous or ichthyoacanthotoxic fish are worldwide in distribution. However, they are found in greatest numbers and in more diversified forms in the torrid zone and most especially in the Indo-Pacific region, from the East Coast of Africa to Australia, the Philippines, and adjacent island groups.

SHARKS, RAYS, AND RATFISH

The important piscine class Chondrichthyes consists of the world's sharks, rays, and less well-known chimaeras. As a class they have well-developed jaws although their skeleton is cartilaginous rather than bony. They are an ancient group of ani-

---

*The venomous nature of the old wife (*Enoplosus armatus*) and of the scats (*Scatophagus argus* and *Selenotoca multifasciata*) has been established.

mals and are represented by some seriously venomous species.

A number of the sharks (of at least 11 genera) have spines in their dorsal fins that could lead one to suspect a system of venom production. Only two species, however, are actually *known* to be venomous. The hornshark (*Heterodontus francisci*) and the spiny dogfish (*Squalus acanthias)* have both been positively identified as venomous. It is likely that at least one other member of each genus will also prove to be venomous once their structures are thoroughly investigated.

The hornshark (also called the bullhead or Port Jackson shark) is a Pacific species (the whole family Heterodontidae is), while the spiny dogfish are found in most seas except the Arctic and

One of two dorsal spines of the spiny dogfish, *Squalus acanthias*. These common coastal fish are involved in accidents with fishermen regularly. They are sublethal albeit painful encounters for the people involved.

Antarctic. Both species apparently migrate in search of suitable water temperature, but their movement is imperfectly understood. Neither is dangerous in the manner usually associated with sharks, that is, neither attacks man as food. But, if either is handled carelessly serious injury can result.

Both the hornsharks and the spiny dogfish have two well-developed dorsal fins, and at the leading edge of each fin a single spine emerges. Seen in cross section these spines are roughly triangular with two sides slightly convex. The third side, the side facing toward the fish's rear, is grooved and it is in this groove that the pearly white, glistening venom-producing tissue lies. Very little is known about the production of venom or about its pharmacology once it has been injected into another organism, but it is known that intense pain results when one of these fish is carelessly handled and a spine encountered.

Dr. H. Muir Evans described a patient brought to him in England with a spiny dogfish injury to his hand. (He referred to the fish as a *spurdog.*) The initial reaction was pain which lasted for almost five hours and was accompanied by considerable swelling. Both wrist and forearm were inflamed and painful. The inflammation began to subside after four days, but it was a week before the swelling and pain in the wrist eased off. At the end of a week the patient was discharged to convalesce at home. (I was somewhat luckier when, in 1968, I carelessly handled a newly caught spiny dogfish. I was jabbed in the palm of the hand, but not envenomated. No symptoms followed except those for a trivial mechanical injury. A certain percentage of dogfish injuries are dry.)

Although shark venoms are believed to be mild when compared to those of some other fish, there have been unsubstantiated reports of human fatalities. If, indeed, there have been some it is at least possible that a severe allergic reaction was involved. This, though, is speculation. (Infection is another and perhaps more likely cause. At least one fatal case of tetanus followed a stingray injury.)

The spiny dogfish showing the position of its two spines, devices to deter predation. Either dorsal spine, anterior or posterior, can inflict a painful injury.

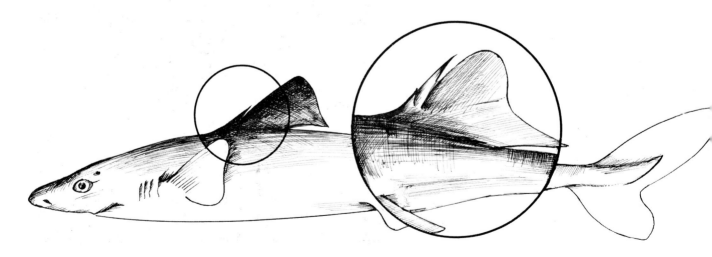

The honeybee, *Apis mellifera,* is not meant to sting and survive. If it uses its sting in defense of self or hive it will perish as a result. The reason for this bizarre development is not known.

The black widow spider, *Latrodectus mactans,* is the world's most seriously venomous spider. Its bite, however, is rarely lethal, though extremely painful.

The red spotted newt, *Diemictylus viridescens*, produces a powerful toxin in its skin that makes it dangerous to attempt to eat. It has no means for using its biotoxin in an attack on another animal, however.

A dendrobatid frog, the red and black, known as an "arrow-poison" frog. This small animal produces one of the most toxic of all known animal substances.

The stingrays are of greater significance than the venomous sharks. They are, in fact, from the point of view of public health, the most significant venomous fish in the world. The marine forms are found around the world in temperate and tropical seas. Venomous fresh-water forms are known to exist in the Atlantic rivers of tropical and temperate South America, Equatorial Africa, and at least one Indo-Chinese river system, the Mekong River of Laos. The list is expected to grow as our knowledge of these animals increases.

Seven families of rays are believed to contain venomous species and the literature that has grown up around them is extensive. This is undoubtedly because more people are injured by stingrays than by *all other fish combined.* In fact the stingrays are one of the three groups of venomous fish that have been studied in any depth at all. (The others are the weever fish and the scorpion fish, particularly the stonefish.)

The seven families believed to have venomous members are the sting or whiprays (Dasyatidae), butterfly rays (Gymnuridae), devil rays or mantas (Mobulidae), eagle or bat rays (Myliobatidae), river rays (Potamotrygonidae), cow-nosed rays (Rhinopteridae), and the round stingrays (Urolophidae). In all families the stinging apparatus is defensive and not used in any other way. The animals themselves are not aggressive and only use their defensive equipment as a reaction to threat or intrusion.

Although rays are sometimes found swimming near the surface (the devil rays usually are), most are typically bottom feeders. They may cruise slowly over sandy bottoms or lie partially buried in the mud. They are difficult to detect and easy to step on, and that is where the trouble starts. Some species are slow to move out of the way even when they sense a disturbance nearby. They apparently depend on their natural camouflage to help them avoid trouble. Like many other animals equipped with a particularly powerful defensive system (the skunks and porcupines are good terrestrial examples), they seem to take it for granted that they will be left in peace.

The venomous apparatus of the ray is contained in its tail. Typically, there is one spine but it is not uncommon to find several in a single animal. Some sources have suggested that the spine is replaced every year, but there seems to be little evidence to support such a claim. It is more likely that they are replaced as they are used.

The spine, a dentinal structure, is enclosed in a integumentary sheath and lies relatively flat in the *cuneiform area,* a depression on the upper or dorsal side of the tail. The spine is erected for use and the entire tail with its powerful muscle structure is used to drive it into an intruder. Tremendous force can be exerted.

Dasyatis americana moving slowly across the bottom. The stinger slightly elevated from the surface of the tail can be clearly seen.

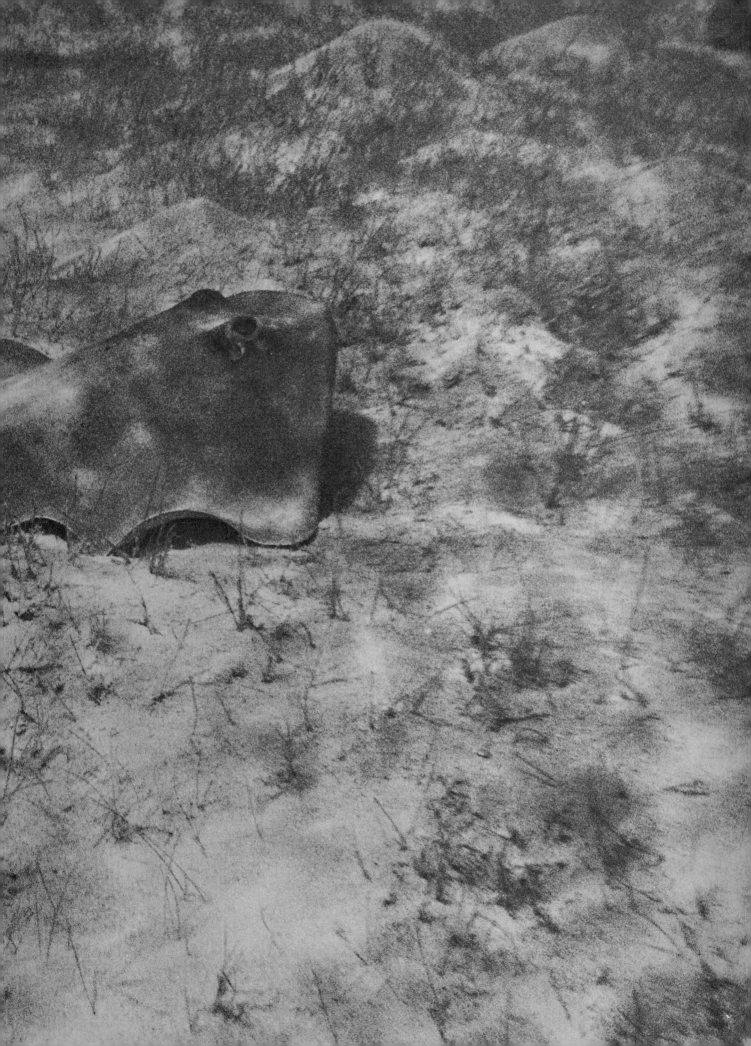

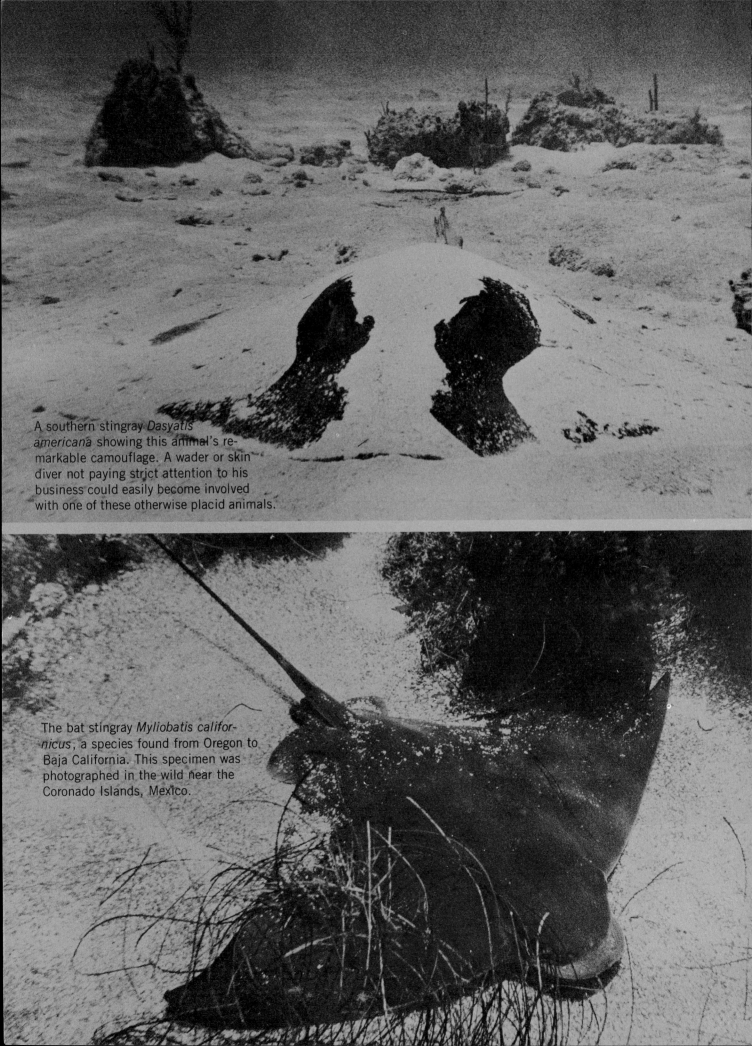

A southern stingray *Dasyatis americana* showing this animal's remarkable camouflage. A wader or skin diver not paying strict attention to his business could easily become involved with one of these otherwise placid animals.

The bat stingray *Myliobatis californicus*, a species found from Oregon to Baja California. This specimen was photographed in the wild near the Coronado Islands, Mexico.

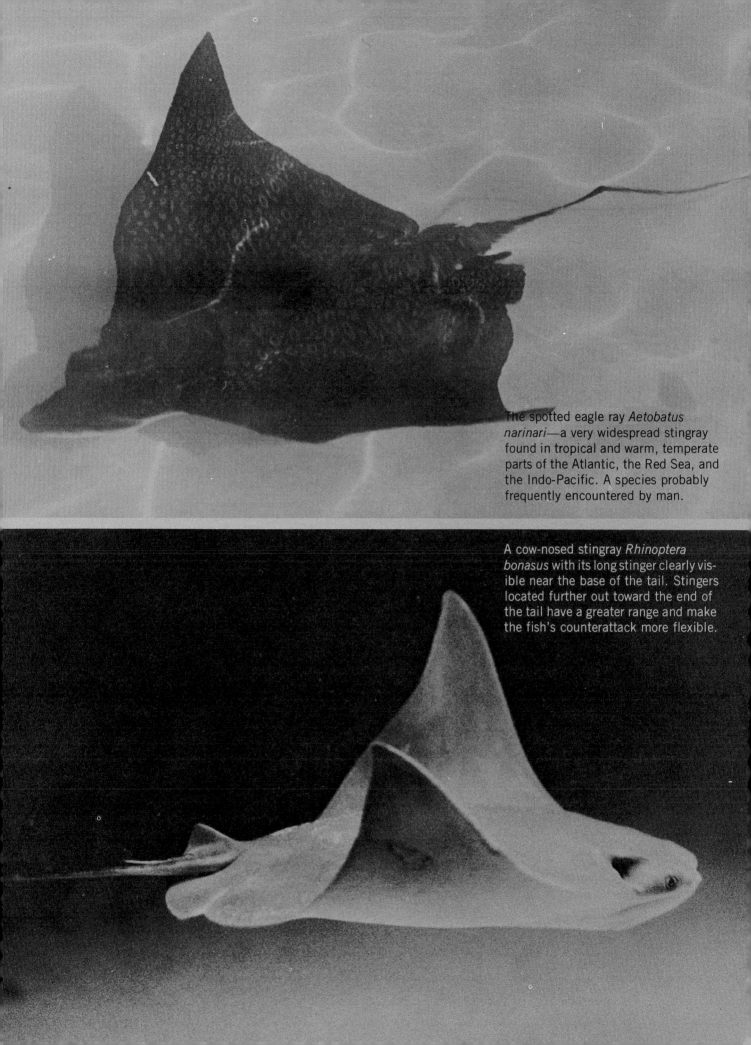

The spotted eagle ray *Aetobatus narinari*—a very widespread stingray found in tropical and warm, temperate parts of the Atlantic, the Red Sea, and the Indo-Pacific. A species probably frequently encountered by man.

A cow-nosed stingray *Rhinoptera bonasus* with its long stinger clearly visible near the base of the tail. Stingers located further out toward the end of the tail have a greater range and make the fish's counterattack more flexible.

As the spine enters a victim the integumentary sheath is ripped away. The upper surface of the spine is marked by a pronounced ridge, and the venom-producing glands lie on either side of this ridge. In addition, there is venom production in the cuneiform area so that even the spine at rest lies in a bed of mucus and venom. The spines themselves are serrated and the teeth serve a double purpose as the device is brought into play. They help tear the sheath that encloses the sting, facilitating the release of venom, and they rupture the tissues around the area of penetration, helping to speed the spread of venom into the victim's blood.

Pain, once again, is the first and often predominant symptom of envenomation. There can be considerable mechanical damage as well and this can lead to severe complications later with infections and even gangrene. Within an hour and a half the pain usually reaches a peak and then begins to subside slowly, but it may be two days before it has completely eased away. There can also be a fall in blood pressure, nausea and vomiting, diarrhea, and other symptoms of acute systemic distress including paralysis and death.

There is no shortage of case histories of stingray victims. Ulysses was said to have been killed by a spear tipped with a stingray spine and Captain John Smith was badly stung while fishing in Chesapeake Bay in 1608. The explorer not only survived; he ate the fish that stung him that same night. Most people are not inclined to do that just after an encounter.

There have been any number of fatal encounters reported, but since it is likely that most stingray injuries never reach the literature it is not possible to estimate what percentage the fatalities represent. (Russell reports two fatal stingray injuries among 1,097 cases in American waters during one five-year period.) But America has good medical facilities and relatively mild stingrays.

Wounds may be either of the puncture type or, as a result of a somewhat glancing blow, of the laceration type. Even when a wound is fairly clean it will be compounded in those cases where the spine breaks off in the wound and an attempt is made to withdraw it, because of its rearward pointing teeth. This problem, however, is usually not encountered.

It is not easy to tell which of the effects of an episode are secondary (the result of pain-induced shock) and which are truly reactions to the venom itself. There is apparently a constriction of blood vessels and a respiratory depression, although all the mechanisms involved have not been worked out.

Dr. Findlay Russell has estimated that there are 750 stingray injuries a year in American waters along and that only 20 percent of the victims are seen by a physician. Six percent, he says,

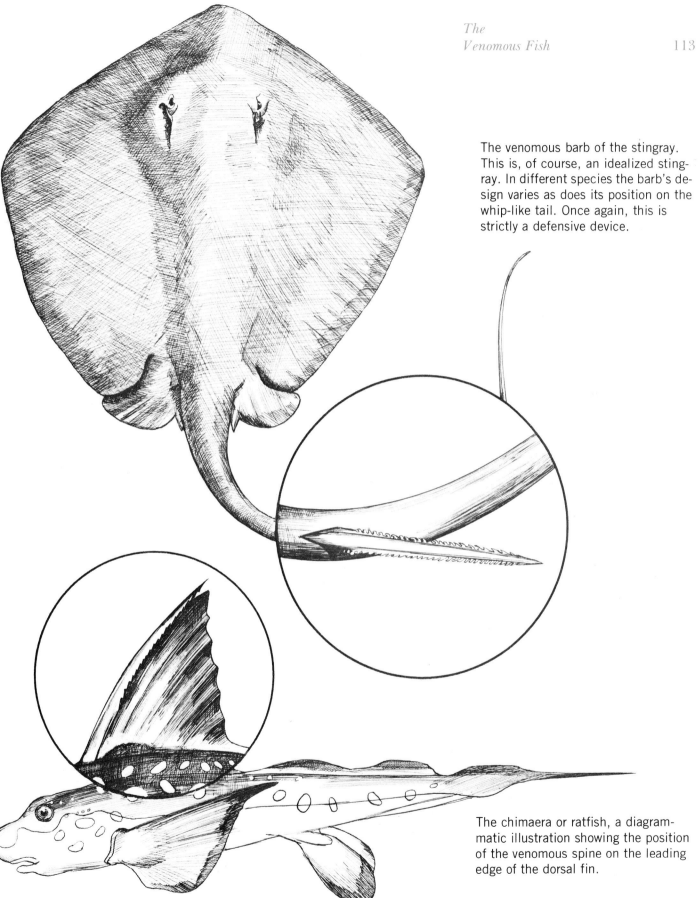

The venomous barb of the stingray. This is, of course, an idealized stingray. In different species the barb's design varies as does its position on the whip-like tail. Once again, this is strictly a defensive device.

The chimaera or ratfish, a diagrammatic illustration showing the position of the venomous spine on the leading edge of the dorsal fin.

are hospitialized. In his opinion, although the principal action of the venom is on the cardiovascular system, there are marked changes in the respiratory and central nervous systems.

The reaction to the sting of a freshwater stingray is apparently similar to that of the marine forms. Dr. Mariano Castex lists among the initial manifestations "acute pain with or without spasms or muscular cramps" and "anguish and restlessness." Among the subsequent complications he lists gangrene and "formation of chronic lesions surrounded by inflammation." He places healing time at "between twenty days and several months."

Clearly, an encounter with a stingray constitutes a medical emergency. On some very large specimens the spine is long enough to cause severe injury without regard to the venom. One victim was stabbed in the liver. Extreme caution is advised in handling specimens inadvertently netted or hooked; similar wariness should be shown by people who wade where stingrays are known to exist. Thousands of people a year can attest to the consequences of lack of caution. More than a few people have not lived to so attest.

The chimaeras, or ratfish, are related to the sharks and rays and two or three of them at least are equipped with a venomous spine, that is, a spine associated with venom-producing tissue. Little is known about the group as a whole and very little work has been done on their venom systems. Those known or at least believed to be truly venomous are a ratfish with rather broad distribution in the Atlantic Ocean—*Chimaera monstrosa*–and another from the Pacific Coast of the United States—*Hydrolagus colliei*. A congener from the North Atlantic—*H. affinis*—is apparently venomous as well.

Primitive, tapering, whip-tailed, weak swimmers, the omnivorous chimaeras have a single dorsal spine as their envenomating device. It is located on the leading edge of a prominent dorsal fin just in back of the head. The spine is rather long and has serrated trailing edges—two corners of a rough triangle. The teeth are more prominent toward the spine's tip and they point downward. The spine is enclosed in an integumentary sheath. Along the rear edge of the spine, where it stands adjacent to the fin, there is a shallow grove filled with a strip of pale tissue, the venom gland. For part of its length the spine is attached to the dorsal fin by a white membrane. This membrane probably produces at least part of the animal's venom.

A rather complex system of musculature enables the fish to erect both the spine and the back fin at will. On contact with the sharp, erect spine envenomation takes place, assuming a wound has been inflicted. The sheath is stripped away, at least in part, and the wound receives venom the nature of which is unknown.

Say's stingray, *Dasyatis sayi* (also called the blunt-nose stingray), is a species of the western Atlantic found from New Jersey to Brazil.

The freshwater stingray *Potamo-trygon motoro* found in the river systems of Argentina, Paraguay, Uruguay, and Brazil. A widespread and potentially troublesome species.

Envenomation results in pain, local symptoms, and, in laboratory mice, death after a whole range of severe systemic reactions. Although a certain number of fishermen are apparently injured by ratfish every year (the chimaeras take almost any bait), no statistics of the injuries are available and claims of human deaths remain unsubstantiated. The unknown aspects of the chimaera life-history and venomous potential are well suited to the creature's name. In mythology *Chimaera* refers to a composite beast: tail of a dragon, body of a goat, and head of a lion. For all of that we do know the chimaera is all fish albeit a singularly unattractive one.

The sharks, rays, and ratfish are extraordinarily ancient creatures. There can be no doubt that they developed their venom as a means of defense. Located as they are well away from the animals' mouths the venomous spines have never had a role in food-getting. And although there is a tendency for men stuck by one of these creatures to take the offense personally, these animals had their envenomating equipment scores of millions of years before man evolved. It is just that we, when we are careless, are as vulnerable to that painful and sometimes even deadly jab as any large predatory fish might be. There is nothing personal about it, except the pain and the sickness.

CATFISH

As a very small boy I remember the great pride I felt in pulling a stout, slimy fish from the river behind our house. It had taken a worm and although it did not offer too much of a fight in the water, it thrashed around most satisfyingly once it was on the bank beside me. Afraid that it would flop right back into the river and be lost (it had thrown the hook as I landed it), I grabbed it with both hands. Triumph turned to tragedy and a very sad little boy limped home weeping, both hands red and swollen. I had grasped a "horn pout" and had learned about venomous spines. (If I had lived in the Midwest instead of New England, my "horn pout" would have been a "madtom.")

The catfish are widely distributed—they are both freshwater and marine—and strangely diverse in body style. They range from very small to very large, from stout to eel-like. What they have in common is a place in the suborder Siluroidei (order Cypriniformes) and venom-equipped spines in the dorsal and pectoral fins. There are about a thousand species of catfish and most of them are freshwater. Our discussion will be generalized since a great deal of data is missing. We will refer to both marine and freshwater species.

There are variations on what might be considered a basic catfish venom apparatus design, and most of these are not clearly understood or at least have not been adequately described.

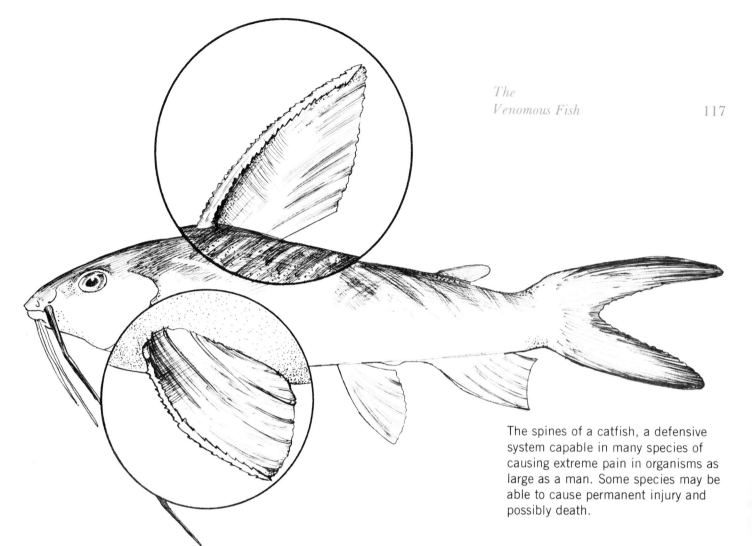

The spines of a catfish, a defensive system capable in many species of causing extreme pain in organisms as large as a man. Some species may be able to cause permanent injury and possibly death.

There is a general form, however. Catfish have dorsal and pectoral spines, parts of fins or rays that have been modified to this special use. They are equipped with venom or axillary glands. There is also a system for erecting or extending the stinging spine. As many as four pairs of muscles may be involved and structurally the whole system is surprisingly complex.

The spines themselves are serrated and are enclosed in the familiar integumentary sheath, the sheath this time being an extension of the fin adjacent to the spine. (At least in some species this is true, and at least of the dorsal spine.) The complex design of the spine's base enables the fish to lock it into position. This ability, of course, facilitates the spine's use since it can only be deflected if the fish's entire body is.

The venom-introducing spines of the catfish have a strange method of growth that adds a peculiar feature to their form. Length is added not by growth from behind, the way a fingernail grows for example, but by a new tip appearing in the form of a cap. Eventually this soft cap (it is often called a *spurious ray)* hardens, forming new serrations and adding new length to the business part of the spine. At that point, of course, a new soft tip has developed. Whenever the sting is employed by the fish the newest section, useless in combat, is demolished to allow the last tip to be employed in penetrating the enemy. This is apparently true of both the pectoral and dorsal spines in (surmise here) perhaps all species that are venomous. The venom glands associated with each spine extend out to the tip, enveloped by the sheath. When the sting is employed the sheath and the glands are ruptured and the venom flows into the wound. Because the fish can lock its spines in place, and because the spines are very sharp, a wound can be either a puncture or a laceration.

As we should have come to expect, an immediate reaction to an encounter with a catfish spine is pain. (I can personally attest to this.) It can be quite intense and depending on species and size and on the depth of the wound and amount of venom received, can last for minutes or days. There are systemic effects and these apparently follow a rather generalized format of intoxication: weakness, nausea, fainting, a very rapid but weak pulse, depression of blood pressure, difficulty in breathing. The young fisherman who encounters a river "cat" and gets stung is likely to compare the experience with a bee sting. That was my case, but it is not satisfactory as an overview. Reportedly, some catfish have killed creatures as large as men. In the family Doradidae there is a fish known as the doradid armored catfish (*Pterodoras granulosus*). Known as *bagre* in Latin America, it is found in rivers feeding into the Amazon and other systems on that approximate latitude. It has reportedly killed.

A catfish from the family Pimelodidae (*Pimelodus clarias*) is found in Central America and perhaps parts of South America. It, too, has reportedly taken human life.

A marine catfish from the Indo-Pacific region (*Plotosus lineatus*)—generally known as the *oriental catfish*–has apparently claimed human life on a number of occasions. It is dreaded wherever it is found and with good reason. It is rather striking as catfish go, brown with yellow stripes that run almost its entire length, but it should be left strictly alone. Catfish can be grasped firmly around the body behind and out of reach of the pectoral and dorsal spines, but they are slimy and difficult to hold. A "horn pout" or bullhead in a Massachusetts stream and the risk of a wasp-like sting is one thing; cold clammy skin, failing respiration, and coma are quite another. *Plotosus lineatus* should not be handled. A specimen inadvertently netted or otherwise caught should be disposed of only by the use of mechanical devices. Handling such a creature is a bad gamble.*

*Reports of human deaths from contact with *Plotosus* are quite old and held suspect by some investigators.

The weeverfishes belong to the family Trachinidae and are among the most dangerous of all venomous fish. They can and do cause human deaths. They are apparently limited in distribution to the Eastern Atlantic and the Mediterranean. At least four species are seriously venomous. They have many common names in the various parts of their range. The name *dragonfish* is generically applied and indicates the special regard in which these fish are held by the fishing peoples who have had to contend with them down through the centuries.

The weeverfish are small animals. The largest, *Trachinus draco*, a fish well-known before the birth of Christ as seriously venomous and equipped with "prickles," barely exceeds 18 inches. The weevers are often sedentary and will lie partially concealed in a sandy bottom. When disturbed, however, they can move rapidly and are apparently aggressive.

Characteristically the weevers have two dorsal fins. The hindmost is much longer and extends almost to the tail. The smaller forward fin contains from five to seven venomous spines. In addition, the weevers have one venomous spine each on their two opercula or gill covers. These, together with the anterior dorsal fin spines, constitute their dangerous and occasionally deadly envenomating apparatus.

One of the world's best known seriously venomous fish, the greater weever *Trachinus draco*. This fish was known to and discussed by the ancients. Although generally sublethal this species can cause human death.

When threatened the weevers may swim at and attempt to strike their enemy with their spines. They apparently employ the two opercula spines in an attack (counterattack, from their point of view) and hold their dorsal spines erect and ready should they be grasped, trod upon, or bumped.

The spines of the weevers are enclosed in a thin integumentary sheath that also encloses the venom glands. These glands, which lie in grooves in the spine, are pale to white, spongy, and run almost the entire length of the spine. The act of envenomation is accomplished in the typical piscine fashion. The intruder is jabbed with a spine, the sheath is ruptured, and venom flows into the wound.

Since weeverfish are common in heavily fished waters, encounters by man are frequent. They are not always discarded, for their flesh is tasty. In some areas laws have been required to keep fish dealers from selling weevers without first removing their venomous spines. Special problem arises from the fact that weevers live a long time out of water. Handled when assumed dead they can inflict a serious wound.

Weevers can and do kill although the typical intoxication is not lethal. Extreme pain is usually the first symptom. Tales of men being driven mad, crying for a merciful bullet, and cutting off their own finger or limb in order to obtain relief are frequently encountered. They are apparently accurate. Shock is a natural concomitance, followed by variously severe local and systemic reactions. There can be swelling and discoloration. A burning or even crushing sensation can accompany the intractable deeper pain. Headache, vomiting, respiratory and cardiac distress, convulsions, and death can ensue. Most encounters are less severe. Although much work has been done in this area the information on the chemistry of weeverfish venom is far from complete. Structurally the venom is complex. Adrenaline is present in considerable quantities as is noradrenaline. There are chemical differences between the venoms of the several species. In all species, however, the venom is potentially destructive to human beings.

SCORPION FISH AND STONEFISH

Tropical and temperate seas are the major scorpion fish areas, although some species apparently extend into Arctic waters. Some of the scorpion fish grow to large size and are used for food while others are commercially worthless because of size or lack of palatability. The taxonomy of the family is open to question, and we shall follow Halstead's lead and think of it as one broad family that includes the stonefish rather than as a proposed grouping of two families, different and distinct.

Halstead lists 57 species as reportedly venomous. It is difficult to make many meaningful generalizations when writing of so many species, or so many forms, with ranges that extend from Norway to New Jersey, from Costa Rica to China, Africa, Brazil, and Australia. This is even more true when dealing with a group of venomous animals with a venom range from mild to deadly. We shall attempt to determine some common facts about the entire family, allowing for the differences that must obviously exist.

There are about 350 species of scorpion fish all told and it is quite possible that many more than the 57 in Halstead's interim listing will eventually prove to be venomous. As a group they are shallow water fishes that live around rocky, coral, or kelp features. Some species lie on sandy bottoms, frequently half buried and all but totally invisible. Others are more active middle water swimmers.

The scorpion fish vary remarkably in appearance. Specimens of the genera *Brachirus* and *Pterois* (variously called zebrafish, lionfish, turkeyfish, butterfly cod, etc.) are among the most flamboyant and stunningly beautiful fish ever discovered, specimens of the genus *Sebastodes* (variously called stingfish and versions of that name) are quite ordinary looking, while specimens of *Synanceja, Scorpaenopsis, Sebastapistes,* and *Scorpaena* are among the wartiest, slimiest, grossest looking creatures in the sea. Seen side by side the "aesthetic" range, while admittedly personally arrived at, is amazing. Nonetheless, they are all scorpion fish and have a most unaesthetic potential.

At least as far back as Aristotle (384—322 *B.C.*) man was writing about the scorpion fish. And that concern has been justified. With the medical attention available even a hundred years ago, envenomation must have been a frightening episode.

The most startling members of the family are the bizarre zebrafish. There are a number of species and they belong to at least two genera, but we can take one of the more seriously venomous for our example—*Pterois.* Like skunks and porcupines, the zebrafishes apparently find it quite impossible to believe that they are vulnerable to attack. They swim around in the open, often slowly, and often in wide, leisurely circles, spreading their grand, lacy fins in all their splendor. Their pectorals, particularly, are beautiful and arresting. But, beneath all of the zebrafish's beauty there lies a deadly serious envenomating apparatus. When disturbed the zebrafish (or turkeyfish as it is often called because of its midwater "strutting") erects its dorsal spines and spins its body so as to meet the intruder with its most deadly equipment. It may advance on and thrust at anything it finds threatening. Zebrafish are not conscious of their power in the sense we would normally associate with that word, but they "know" in their own way that they are able to fend off and cripple or destroy almost anything that would dare disturb them.

The fiercely ugly *Scorpaenopsis,* although not given to swimming about in middle water, will come at an intruder and attempt to jab it with venomous spines. The rushes are short and clearly defensive, but to an unwary wader the result can be agonizingly painful or worse.

In contrast, the most venomous species of all, the stonefish, seldom make any move toward an enemy. They remain hidden in perfect camouflage among bottom rubble, warty, slimy, sometimes all but covered with colonies of algae, and wait to be bumped or stepped on. As an enemy approaches they elevate their venomous spines to give maximum cover, but the rush and jab is generally not attempted. In a way this stationary holding habit makes them even more dangerous. They do not give themselves away and can be all but impossible to detect and avoid.

The venom apparatus of the scorpion fish is what we have come to see as typically piscine. It includes a spine, grooves in which the venom-producing gland lies, and integumentary sheath that encloses the whole apparatus. The sheath is stripped away as the spine enters the victim's tissues and the venom flows. In most scorpion fish the venom glands are long and slender and are not connected to a duct of any kind. In the stonefish *(Synanceja)* the system has been more perfectly refined. There is a large, bulging venom gland a little further than midway up from the base and an actual duct to conduct the poison into the wound. Very little is known about the venom but its dangerous fractions are believed to be protein in nature.

The zebrafish has a large number of venomous spines to array around itself when in a defensive attitude. There are 13 dorsal spines, three anal spines, and two pelvic spines, one on each side. An encounter with any one of the 18 can cause extreme distress and even death.

The scorpion fish *Scorpaena* (there are at least 15 forms in this genus known or believed to be seriously venomous) has only one less dorsal spine—12 instead of 13. It has the same armament of three anal spines and two pelvic spines. The spines tend to be heavier and somewhat shorter than those of the zebrafish and the sheath that covers them is thicker. The other differences, although marked and apparent to the expert, are technical matters.

The really terrifying array of the stonefish numbers 13 dorsals, three anals, and two pelvics. But the stonefish's spines are heavy and the sheath gross, thick, and warty. A well-developed venom duct leads up from the enlarged mass of venom-producing tissue.

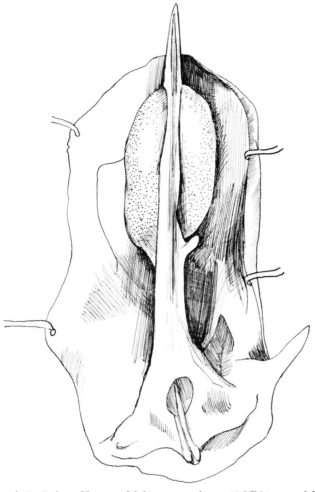

The deadly venomous spine of the
stonefish dissected to show the location of the venom glands near the tip.
When the sheath surrounding this
spine is torn away, these glands rupture and the venom flows into the
wound.

Richard C. Schaeffer and his co-workers (1971) provide a general description of scorpionfish venom: "The stings of these fish
lead to pain, localized swelling, discoloration and paresthesis
around the wound, and in some cases to lymphadenitis, lym-
phadenopathy, nausea, vomiting, weakness, pallor and syncope.
In the more severe cases there may be intense pain, respiratory
distress, shock, coma and death."

The Commonwealth Serum Laboratories in Melbourne, Australia, produce the only piscine antivenene now available. It is
specific to the stonefish and an instructional leaflet that is supplied with the antivenene remarks:

> "The dominant symptom of Stone-fish sting is
> agonizing and persisting pain. Oedema, which usu-
> ally develops rapidly after a sting, may become ex-
> treme. Abscess formation, necrosis and gangrene
> have occurred in untreated cases. In addition to the
> local effects, muscle weakness and paralysis may de-
> velop in the affected limb and varying degrees of
> shock may occur. The systemic effects are due to the
> presence of potent myotoxins, which act directly on
> all types of muscle—skeletal, involuntary and car-
> diac.

A stonefish *Synanceja horrida* showing the foremost dorsal spine exposed. This is almost certainly the most venomous of all living fish now known to science.

On the subject of pain, people stung by stonefish have been known to plunge the affected member into an open fire and to attempt self-inflicted amputations. The pain is apparently unbearable.

Although edible, the stonefish is soundly despised over its vast tropical Indo-Pacific range. The three most commonly encountered forms (*Synanceja verrucosa, trachynis,* and *horrida*) all lurk ominously in intertidal zones offering excruciating agony to anyone bumping or stepping on them. The occasional reports of a trivial wound, one without envenomation, are probably either cases of mistaken identity or of lucky chance encounters with an incomplete spine—one recently discharged or underdeveloped. They are known to occur.

The South African ichthyologist J. L. B. Smith has provided case histories of two fatal encounters with stonefish. A fifteen-year-old boy whom Smith described as "athletic and well-built" received three stonefish stab wounds in the sole of one foot while wading in shallow water near Mahé Island in the Seychelles. He was in agony almost immediately and nearly collapsed. Then he began turning blue and was seen to froth at the mouth. He died in a car on the way to the hospital. Death was listed as a result of cardiac or respiratory failure.

In the same year (1956) near Pinda, Mozambique, two men were spearfishing near an offshore ridge at low tide. They were about five miles from shore when one noticed that his companion had collapsed. By the time he reached his friend the latter was delirious but did manage to say that he had been stabbed by a "sherowa" or stonefish. He was dead within the hour. Smith later examined the body and determined that there had been a single stab wound in front of the second toe of the right foot.

The bullrout *Notesthes robusta*, a small Australian coastal fish with venomous spines on each side of the head in front of the gill covers. It has a sublethal but very painful venom. It is related to the stonefish but is not nearly as dangerous.

The wound was three-fourths of an inch deep. There was no apparent discoloration or swelling. The man was between thirty and forty years old, strong, and reportedly in good health.

It would be an exaggeration to say that these encounters were typical, but unfortunately they cannot be considered either rare or unexpected. The stonefish is universally dreaded. The Australian Aborigines create a beeswax model of the fish and act out the effect of an encounter in elaborate pantomime. The purpose, apparently, is to educate the young.

A survey of stonefish injuries that was done in 1889 listed seven known fatalities out of 25 cases. The time of death varied in the accumulated reports from a few hours to several months. There can be little doubt that many, if not all, of the deaths that occurred after the first few days would have been avoided had modern medicine been available. There are very serious subsequent effects in untreated or improperly treated stonefish injuries.

Fortunately, there is an antivenene available as well as ways to relieve the excruciating pain resulting from stonefish envenomation. (There can be no doubt that extreme pain can bring on or increase the severity of shock, and shock alone can kill.) Whereas morphine has virtually no effect on stonefish victims, emetine hydrochloride injected into a wound brings almost immediate relief. The problem, of course, is to have emetine hydrochloride at hand and antivenene quickly available. Since the stonefish ranges from the East Coast of Africa, through the Red Sea and the Indian Ocean to Australia and in the Pacific to the Philippines and Tuamotu Islands, it is impossible to have these medicines at hand for the vast majority of potential victims. Many people who encounter these fish, even in our own time, never see a doctor, much less sophisticated medication.

As for the toxicity of stonefish venom, each of the two sacs found on each of the 18 stonefish spines contains ten milligrams of venom—one researcher (Atz) has provided for comparison the information that a postage stamp weighs 65 milligrams. However, those ten milligrams are enough to kill 1,000 mice. If all spines and all venom glands were equally well developed (this is by no means always the case), a single stonefish less than a foot long would be equipped with enough venom to dispatch 36,000 mice. What precisely, one wonders, is the stonefish equipped to defend itself against? The fish is well defended by its appearance and habits. Its ability to remain virtually motionless for an almost interminable length of time and the fact that it looks so much like coral rubble that it must rate as one of the best camouflaged creatures on earth would seem to afford protection enough. The spines and their venom are never used in food-getting for here again camouflage is the fish's device. It lies perfectly still until a likely meal drifts or swims innocently by. Then, explosively, the hump of slimy, algae-covered bottom rubble comes to life. A gaping maw in a grotesquely oversized head envelops the prey as the stonefish snaps closed and returns to its camouflaged position. Why, we must ask, why venom in such quantities and why a venom delivery system, including clearly defined ducts, so far advanced that it surpasses in sophistication the envenomating apparatus of almost all other fish? There does not seem to be a logical answer except, perhaps, to note that man apparently did not invent overkill.

TOADFISH

The family Batrachoididae contains at least 15 species of repulsive little toadfish that are believed to be venomous. They are bottom fish and largely marine although some few are estuarine and even ascend rivers. While they are little known in comparison with the stingrays and stonefish, the toadfish are distinctly venomous. Yet no human fatalities resulting from their envenomation have apparently been recorded.

The toadfish's venom equipment differs from that of the other fish we have been discussing in that the glands producing the toxin do not lie along the spine or cluster halfway along its central ridge. The spines are hollow in the manner of a snake's fangs and the mass of pulpy white tissue that produces the venom is at the base of the spine. There are two such spines on the fish's back, variously a quarter to a third of the animal's length back from its head, and one spine on each gill cover. The system, therefore, is distinctly different from that of any of the other fish we have discussed.

A toadfish showing the location of the venomous spines. These devices play no part in food-getting. They are defensive only.

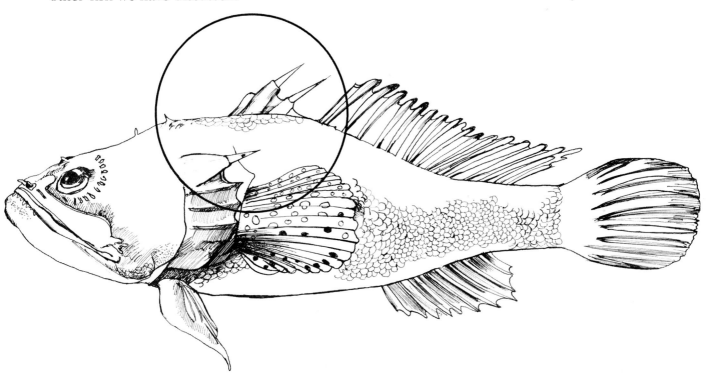

The two dorsal spines are enclosed in a single sheath but each has its own venom supply. They are erected as a single unit when the fish assumes a defensive position and they can inflict a very painful wound. Most wounds are the result of a victim's stepping on a fish although occasionally the opercula spines are brought into play when a fish is being carelessly handled. A clear, slightly acid venom can be obtained from specimens by squeezing the base of the spine (this is not a recommended activity), and experiments have provided a preliminary opinion that the substance is seriously neurotoxic. Little else is known about it.

OTHER VENOMOUS FISH

As was indicated at the beginning of this chapter, there are many fishes in addition to those we have discussed that have been accused of being venomous. Some of them clearly are. The several stargazers (family Uraniscopidae), small bottom marine fish, are equipped with two *cleithral* or "shoulder" spines and these are enclosed for most of their length by a true venom gland. Carelessly handled stargazers can give a painful sting. They are said to have caused human deaths, but very little else is known about the medical aspects of stargazer envenomation.

At least some of the eight rabbitfish (all genus *Siganus*), tropical marine fish found from the Red Sea to Polynesia, are certainly venomous as are some of the surgeonfish (family Acanthyridae), although there are probably no really seriously venomous species in the latter group. There are a number of other venomous fish groups as well, but little is known about them.

The ability of some fish to array a series of painfully or even lethally venomous spines in their own defense is widespread. As far as we know this ability is always and exclusively defensive in use and, unless reports concerning certain moray eels (that seem questionable at best) are substantiated, we can say that no fish has a venomous bite. (I believe that to be the case.) We will be seeing how different is the case with reptiles.

ICHTHYOCRINOTOXIC FISHES

At least 35 species of fish from five different orders (and ten different families) have been reported to be *ichthyocrinotoxic*. That formidable word means, simply, that they are capable of producing a venom, but have no means of injecting it into an enemy. It is apparently discharged into the water as a repellent. Little is known about the venom and there is a temptation to liken it to the venomous skin secretions of certain amphibians. The ichthyocrinotoxic fishes present an interesting area for study, certainly, and someday be shown to play ecological roles now barely suspected.

This concludes our brief review of the venomous fish. Many, many more fish are poisonous to eat than are venomous to touch, and the former are of far greater importance since so many of them are consumed each year. Still, there is a kind of fascination in contemplating bizarre creatures such as the zebrafish, the giant stingray (the Australian giant, *Dasyatis brevicaudata*, attains a length of over 14 feet, a width of over seven feet, and a weight of almost 750 pounds), and the stonefish. Somehow they are unreal, nightmare creatures that can take life by simply brushing against it. But, they do indeed exist. They are among the most formidable animals on earth, and among the strangest.

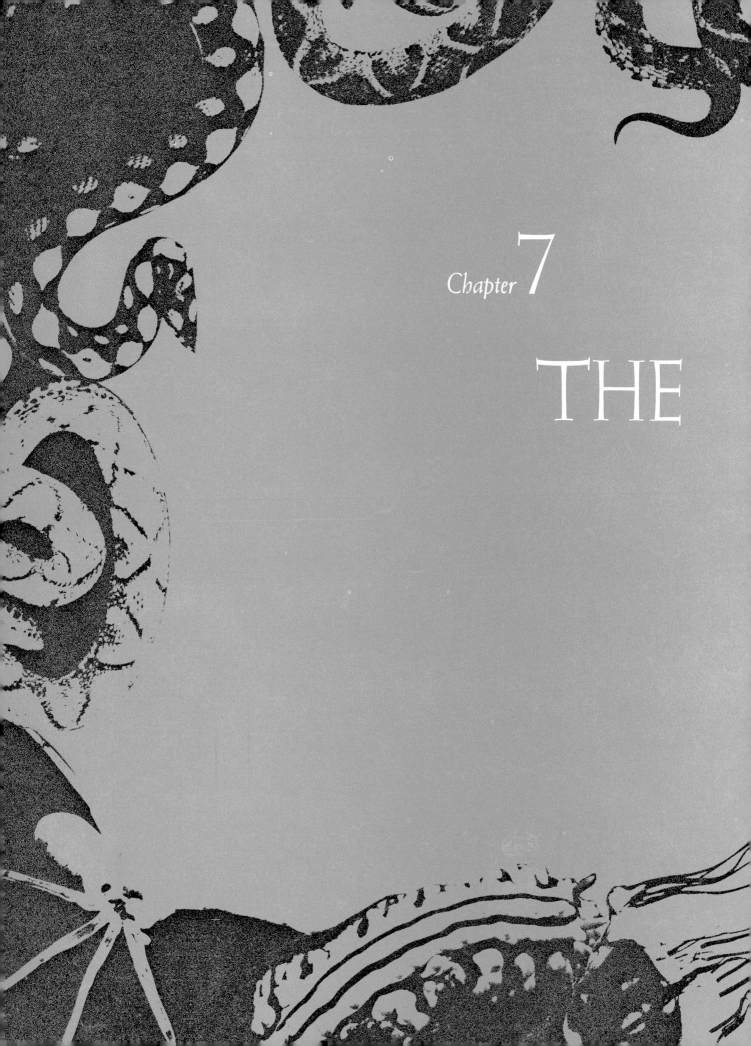

Chapter 7

THE

AMPHIBIANS

The arrow-poison frog, *Dendrobates auratus*, a small animal that secretes one of the most powerful biotoxins presently known. It has no way of projecting its venom onto another organism.

he amphibians are of special interest because of their place in evolution. They were the first vertebrate animals to survive in the hostile land environment and formed the link between fish and reptile. When they crawled out of that ancient sea, these animals brought our potential with them.

Amphibia is a small class. Among its principal orders are:

The Apoda, or caecilians, number no more than 75 species and are limbless and almost totally subterranean. (One genus, *Typhlonectes*, is aquatic.)

The Caudata, or salamanders, are generally restricted to the Northern Hemisphere. The actual number of species is open to some debate but 280 is a safe round figure.

The Anura is the most widespread, familiar, and populous order, the true toads and frogs. They are a group of great interest to us. There are probably over 2,000 species.

The amphibians generally are inoffensive animals. Although a few grow to fair size and a very few can do some mechanical damage by biting, they are the exceptions. The amphibians in general have developed another and quite remarkable system of defense. It involves some of the most powerful and complex biotoxins known.

The biotoxins of the amphibians are defensive substances, clearly that, but the animals producing them cannot bring them into play unless an aggressor comes in contact with them. These substances are too powerful, however, and too clearly intended to be inflicted upon other animals for us to overlook them, even if they are not part of an *active* envenomating system like those found in the other animals we have been discussing.

Although the internal tissues of some amphibians, the organs, and the blood, and the eggs of others are poisonous, the toxins of these animals are usually concentrated in their skin. They have no scales, no feathers, no fur. Their skin must be kept moist, or slimy, so that it is permeable to gases, since it is instrumental in the respiratory process. The skin houses the glands that produce the venoms.

FROGS AND TOADS

In the frogs and toads there are two types of glands. The *mucous glands* are the smaller and shallower of the two. Although generally not as localized as the *granular glands*, in some instances they are more plentiful on some parts of the body than on others. Generally these are areas most likely to be seized by a predator. The secretions from the mucous glands are very fluid, and often slimy. Their primary function may be to keep the animal moist and protect it from desiccation. The granular glands secrete a thicker, creamier substance. They tend to be clustered, often on the sides of the head and on the shoulders. These are believed to be primarily venom glands.

Typical of the granular glands are the parotoid glands of the *Bufo* toads. (The name *parotoid* is suspect since the glands referred to in no way resemble the mammalian parotid glands, which are salivary glands located in front of the ear.) In the *Bufo* toads the parotoid glands are clustered behind the eyes, on the sides of the head, and on the shoulders. The shape and size of these clusters, and their distribution, vary and often can be used to identify species. The individual glands are bulging, granular, and secrete a creamy poisonous substance.

Some species in the genus have additional glands located on or about the limbs. Once again, they are situated in places where the animal is most apt to be grabbed by a predator. Both the head and the hind limbs are likely places for such an encounter. These secretions may also have another use. They may retard the growth of microorganisms. It is quite possible that small moist amphibians could provide such an ideal culture medium that they could be overwhelmed and suffocated by minute plants and colonial animals if they did not have a means of repelling them. This function of amphibian skin toxins is very imperfectly understood.

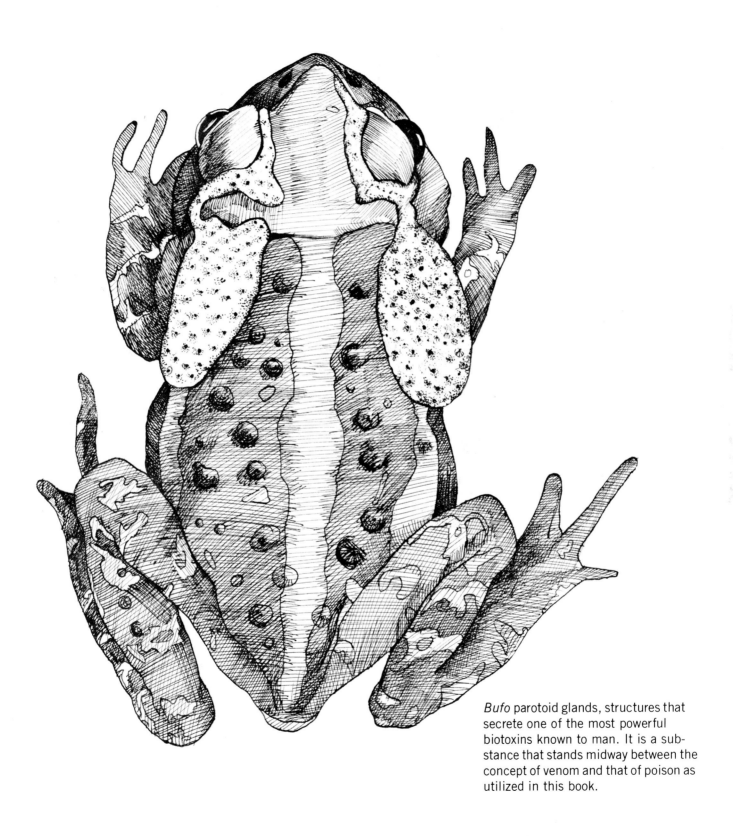

Bufo parotoid glands, structures that secrete one of the most powerful biotoxins known to man. It is a substance that stands midway between the concept of venom and that of poison as utilized in this book.

Bufo marinus, the marine toad. Glands strategically located on this animal's skin secrete a powerful and repugnant toxin into the mouth of any would-be predator.

Toad venom, a purely defensive substance, would kill almost any attacker outright if the toad had a means of introducing it into its foe's bloodstream. But, this, obviously, the toad cannot do. It must depend on the attack, the open mouth, the vulnerable mucous membrane that lines the mouth, and the quick withdrawal in haste and disgust. (When marine toads were introduced into areas like Oahu and Miami, the local dog and cat population suffered heavy casualties until the survivors learned to let the big toads strictly alone.)

Toad venom is extraordinarily complex, and unique components are revealed by the different species. Many of these have been subjected to analysis and the literature on the subject is now vast, but comprehensible to chemists and few others. The components are so specific in many cases that they can be used in a kind of chemical taxonomy—the presence of a factor being diagnostic as to species.

What is significant is the activity of these venom substances. Some of them are widespread, some of them are unique and indeed were unknown as chemical structures until discovered in various bufotoxins (and that is most significant). Their effects range from uncomfortable and mildly annoying to deadly. They are extraordinarily cardioactive and can cause death by disrupting the normal heart rhythms. Some investigators have suggested that the toxins identified in the skin of some toads and frogs are the most active venoms ever discovered. They were referring especially to a group of South American species commonly called the arrow-poison frogs. The venom of the Colombian arrow-poison frog (*Phyllobates aurotaenia*), which contains at least three major alkaloids, may be "the most active cardiotoxin known."

Another of the Colombian poison frogs—*Phyllobates bicolor*—produces a poison known in the Noanama dialect of the Cholo Indians as *kokoi*. This tiny frog (three-quarters of an inch long and a third of an ounce in weight) is black with yellow strips along its back. It is a handsome little creature. *Kokoi* does not affect human skin when it is intact, but the slightest touch of this animal in an area where there is an abrasion or scratch produces the sensation of a very severe bee sting.*

The Cholos capture the little *kokoi* frogs and impale them on sticks. Roasting them alive over an open fire, the hunters dip their arrows into the skin secretions seeping from the bubbles and blisters that form. One tiny frog produces enough *kokoi* to poison 50 arrows. Used in hunting swift animals like monkeys and birds, the poison darts cause almost immediate paralysis and death. The poison apparently affects the muscles directly as well as the nervous system. It is believed that *kokoi* creates an irreversible block of the neuromuscular transmission, making it not unlike curare in effect.

It is quite amazing how animals differ in their tolerance to the various toad and frog venoms. *Kokoi* is not very effective against other frogs and toads and although mice die from it, they have

---

*When Sherman Minton was collecting tree frogs (*Phrynohyas*) in Mexico, he had his driver help him. As they drove along one night the driver absent-mindedly rubbed his eyes with his frog-contaminated fingers. The pain was immediate and so intense he nearly lost control of the car.

better resistance than either rabbits or dogs many, many times their weight.

In 1960 a series of experiments were carried out using the parotoid gland venom from the marine toad (*Bufo marinus*). The extract was administered orally to a variety of snakes. The snakes were given electrocardiograms at set intervals. At a dosage of 3 milligrams (.0001 ounces) of venom for each gram of snake body-weight (3 mg./g. is considerably more venom than a snake would get from eating a toad) the normal toad-eating snakes showed very little reaction. The non-toad-eaters, however, showed severe reactions and a number died following a lethal cardiac response. When the doses were increased to 10 and 20 mg./g. all the snakes died. Even the toad-eaters could not handle that assault on their hearts.

The interesting problem, though, is not how much it took to kill all the snakes, but the system used by the toad-eaters to protect themselves at the lower dosages. What does one snake have that another lacks which enables it to survive an obviously lethal dose of cardiotoxin? Is it a chemical mechanism in the heart itself or is it something in the digestive tract? The answer is of considerable interest to medicine. (Toad skin is widely used in Chinese folk medicine.) In fact the whole subject of frog and toad venoms and their interaction with other organisms is under intensive study. (The ability of hognose snakes and some of their relatives to eat toads is said to correlate with these snakes' proportionally large adrenal glands.) How all this works is not known.

Before going on to some other amphibians it is interesting to note a peculiarity of toad venom production. It is extraordinarily *slow*. The parotoid glands of a marine toad were squeezed and very nearly emptied. Eleven weeks later they were only two-thirds restored. Other investigators reported four to six weeks for recovery in some species. Evidently, in facing an attack by a predator, very little venom is secreted. Very little, apparently, is all that is required.*

SALAMANDERS

Salamanders are also highly toxic to predators. Again, the defensive mechanism lies in the animals' skin secretions. Interestingly enough, in many species that make the classical amphibian transition from aquatic to terrestrial form, the skin of the younger animal does not seem to be poisonous. But, as the animal undergoes its transformation and emerges upon the land, two changes occur in the skin. The outer layers (dead cells) be-

---

*A number of collectors have reported violent allergic reactions to the mere presence of some species. A collector's bagful of spade-foot toads from Texas reputedly gives one a hay-fever-like reaction in a closed car.

come very hard and leatherlike. This *stratum corneum* serves to protect the living layers of cells below from the desiccating effect of air. The second change is the appearance of large, multicellular poison glands.

Long before the salamander actually undergoes its metamorphosis, plug-like structures start growing in the skin. They grow downward into the layers of tissue below the dermis. Eventually they develop into large round glands with narrow ducts that lead to the surface of the skin. The glands are of two types:

> *(1.) Mucous glands*, the smaller of the two, secrete a clear, thin fluid. The fluid is described as an apocrine, that is, it is secreted by the inner wall surfaces of the glands. These glands are well distributed over most of the animal's body and serve to keep its skin moist and elastic.

> *(2.) Poison glands*, the larger of the two types, are usually concentrated on certain parts of the body, especially the tail. Very often there is one or more rows of them down the animal's back. Their secretion is whitish instead of clear, and sticky and slimy instead of fluid. Interestingly, in some species venom can be squirted from the glandular ducts. James Oliver reported an instance where a northern Pacific red salamander (*Ensatina eschschaltzi oregonensis*) actually drove off an attacking snake by lashing at it with its tail. The upper surface of that species' tail is richly endowed with venom glands.

The secretions of the salamanders' venom glands range from irritating to highly toxic. They exhibit the two primary amphibian venom types. There is the *epinephrine-like* which works like adrenaline, accelerating both heart and respiration rate, and there is the *digitalis-like* which retards breathing and diminishes heart rate while strengthening the beat.

One form of salamander poison, tetrodotoxin, is present in two unrelated animals—a puffer fish and a newt. The toxin blocks the conduction of nerve signals. Found in newts of western North America, it is concentrated in the animals' skin, muscle, and blood. Tetrodotoxin is obviously defensive like all of the other amphibian toxins, but it is not clear why it is found in just one suborder of fishes and one family of amphibians. The toxin is found in the newt's eggs as well as in the adult animal, and the form extracted from the eggs is so highly toxic that one three-thousandth (0.003) of an ounce can kill 7,000 mice. Somehow, by mechanisms not yet revealed to us, newts are immune to their own poison. The fact that 25,000 times as much toxin is needed to kill a frog as a mouse leaves a newt unconcerned.

Animals may differ in tolerance to newt toxins according to age. A trout seven or eight inches long will die within five to ten minutes when fed a single spotted newt (*Notophthalmus viridescens viridescens*). A trout ten or 11 inches long will live as long as half an hour. They all eventually do die, however.

Skin from the eft and adult forms of the spotted newt were variously force-fed and injected into a variety of experimental animals. White mice gasped and gaped and exhibited muscular weakness. Before dying they had convulsions and became paralyzed. Their hearts continued to beat strongly long after they had stopped breathing. The back skin of the eft and the adult newt was found to be more toxic than the belly skin. Toads and snakes were much more resistant than mice but varied among themselves as to tolerance.

The adult rough-skinned newt (*Taricha granulosa*) is another very well-equipped animal. It takes only 0.0002 cubic centimeters of its back skin to kill a mouse within ten minutes. Snakes other than garter snakes are apparently 200 times more resistant than mice while garter snakes, regular newt-eaters, are 2,000 times more resistant. What is the mechanism that protects these animals? We do not know.

The classical symptoms of poisoning by a salamander are restlessness followed by epileptic-like convulsions. The pupils dilate and the reflexes grow steadily fainter before vanishing altogether. Respiration is weak and heart action becomes irregular. Very quickly paralysis, particularly of the hind legs, sets in. Post-mortem results show hemorrhage in the lungs and distension of the veins in the heart, brain, and liver. The toxins of some species affect the central nervous system.

IMMUNITY MECHANISMS

Of extraordinary interest is this matter of resistance. A dog weighing 50 to 100 pounds may pick up a toad in its mouth, drop it and reel away salivating and whining pitifully. The same toad may be taken moments later by a snake weighing less than a pound which will not be bothered by the protective secretions at all. Another species of snake, however, might be killed attempting to swallow the amphibian.

It is not difficult to understand why these immunity mechanisms came into being. They allow predators to utilize as food animals that would otherwise be deadly even though passive. What we do not understand is how these mechanisms work or why they have not appeared to counter other venomous substances with which the nonvenomous members of the animal kingdom have had to coexist. It would certainly be no less a survival mechanism for a cottontail rabbit to be immune to rattlesnake venom than for a garter snake to be tolerant of newt poison. Why one and not the other? Could it be that predators that prey on venomous species tend to develop immunity, while prey subject to attack by venomous animals do not? I do not know this to be so, although I do not know of a single prey animal immune to the venom of its natural predator. Recall the case of the gastropod-eating cone shells. Both prey and predator may be seriously venomous. As far as we know the prey is not immune to the venom of the predator, while the reverse may be true. Here is another area requiring research!

Chapter 8

THE

VENOMOUS
LIZARDS

The Gila monster (*Heloderma suspectum*), one of the two species of venomous lizards on earth. An almost mythical creature, as witness the inappropriate word ''monster'' in its name.

pproximately 3,000 species of lizards have survived into our time and only two of these are venomous, both members of the family Helodermatidae. At one time a third species, the extremely rare and almost unknown earless monitor of Borneo (*Lanthanotus borneensis*) was also considered a member of this family. Recently, however, it has been given its own family, Lanthanotidae. As the original assignment to the Helodermatidae would indicate, the earless monitor is closely related to the truly venomous lizards. The animal lives in remote regions of Sarawak on Borneo's northwest coast. The natives there claim it is deadly but examination of the few specimens held by museums show no indication of a venom apparatus or specialized teeth.

The two members of the Helodermatidae are the Gila monster (*Heloderma suspectum suspectum* and *H.s. cinctum*) and the Mexican beaded lizard (*H. horridum horridum*, *H. h. exasperatum*, and *H. h. alvarezi*). The former is limited in range to Arizona, Utah, a bit of southwest New Mexico, Nevada, northern Mexico (Sonora), and a corner of southeastern California. The Mexican beaded lizard ranges (three subspecies) down the west coast of Mexico and nowhere approaches the American frontier. A fossil member of the family, *H. matthewi*, has been identified in Oligocene strata in northeastern Colorado.

The Helodermatidae probably go back nearly 50 million years. The saurians, or lizards in general, are not very ancient and probably originated during the late Triassic Period about 170 million years ago. That is not ancient as reptiles go, for by that time the dinosaurs were already in a high state of development. The heloderms, the monitors, and that rather mysterious earless monitor from Borneo are either the basic stock from which the snakes eventually arose or at least have common ancestry with the snakes.

Of all the living reptiles the lizards are the most varied in structure, in form, and in function. And of all the strange adaptations the lizards have undergone none is more unusual or more bizarre than the venomous capabilities of the family Helodermatidae.

GILA MONSTER AND MEXICAN BEADED LIZARD

The Gila monster is a member of the Great Sonoran Desert fauna, as its range would imply. The Mexican beaded lizard along the west coast of Mexico inhabits a comparable zone.* Both lizards are large (one subspecies of the Mexican beaded lizard, *H. h. exasperatum*, can grow to a length of nearly 800 mm.), fat when in good condition, and relatively slow-moving. In the warm country they inhabit they apparently live largely on small mammals, nestling birds, and apparently some eggs—both bird and reptile. There seems to be little evidence that claims of their living on insects, arachnids, amphibians, and worms have much to support them. Since the prey of the two species normally consists of helpless mammals, infant rats, mice, and rabbits, and equally helpless birds, one can only wonder why they were given a venom capable of killing a 200-pound man.

The bite of the Gila monster has almost inevitably been described as bulldog-like by anyone who has experienced it. In laboratory experiments, in zoo feeding episodes, both the Gila monster and the Mexican beaded lizard have shown enough power in their broad jaws to *crush* full-grown hamsters to death. Why, indeed, does such an animal require an extremely toxic venom and a system of envenomation that although diffuse when compared with that of most snakes is still highly efficient? Most students of these animals seem to agree that the venomous bite of the heloderms is primarily a defensive device. It may be used to some degree in food-getting, although that is doubtful

*Private correspondence from Sherman Minton reports the Mexican beaded lizard from areas of Colima that are not semiarid. Dr. Minton compared the area to the western Ozarks in vegetation and rainfall. He suggests that the black subspecies (*alvarezi*) may live in even more humid country.

and would be very much a secondary use. The Helodermatidae are among the very, very few animals on earth that seem to have developed a venomous *bite* primarily as a means of defense.

There has been a great deal of nonsense written about the heloderms and, since they are strange even for lizards, that is understandable. Following a very ancient pattern of assigning venom production to different parts of the animal, some early writers suggested that the venom was produced in the lizard's tail. Such comments no doubt reflect folklore, but as recently as 1952 I had this sworn to me as proven fact by an apparently well-educated man. In fact, the Heloderms store food in their chubby tails. When in good condition these lizards have very round and full tails, only a very small percentage of which are bone or sinew. In animals in poor condition the tail is characteristically much smaller with its store of fat depleted. The delightful absurdity about venom production in that area has it that the lizard is incapable of eliminating metabolic wastes. The material that would otherwise pass as fecal matter accumulates in the tail where it understandably putrifies. By a means not very well described, this putrifaction is said to pass the full length of the animal's body and be present in its mouth when it bites. It is little wonder, therefore, that for over 400 years it has been said that the Gila monster and the Mexican beaded lizard have venomous breath! With that kind of exudate coming forward the lizard's breath would certainly be bad if not deadly. The early Spanish reports of venomous breath may have been picked up as ancient beliefs of the Indians they encountered in Mexico. At present we do not know if that was so.

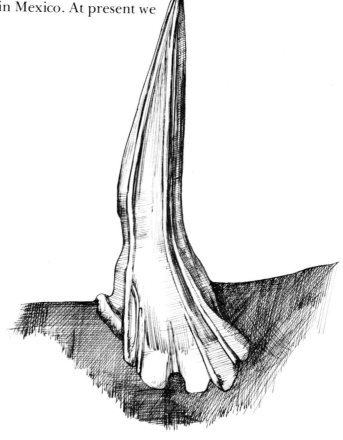

The fang of the Gila monster showing its grooves and cutting ridges. It is a highly specialized tooth evolved to facilitate the flow of the lizard's venom into well-blooded tissue.

All legends aside, there is quite enough truth to make the Gila monster and its Mexican co-species fascinating as well as potentially dangerous animals. The danger from these animals, it should be stressed, can only arise through human carelessness. Gila monsters, although not particularly pleasant, are not aggressive and certainly not so toward anything as incredibly formidable as a man. When unable to retreat, which it almost inevitably tries to do, the Gila monster will stand its ground, hissing and snapping at the intruder. Because it must truly bite and cannot stab forward as a snake does, it is almost impossible for the lizard to grab onto a person unless a hand is offered. Almost all Gila monster bites of record have been from captive specimens carelessly handled.

The venom of the Gila monster does not chemically resemble snake venom. Since it is not part of an essentially food-getting apparatus it is poor in enzymes, except for hyaluronidase. Were the venom an essential or even an important element in the Gila monster's feeding pattern, it would contain digestive enzymes to help the lizard process the prey which, like that of the snake, is swallowed whole. The principal toxic element of heloderm venom appears to be a polypeptide rather than a protein.

The venom arises in modified salivary glands—submandibles—toward the rear of the lower jaw. Here is one way in which the venomous lizards differ from the venomous snakes. In the snakes the glands that produce the venom are *above* the upper jaw. In the lizards they are *in* the lower jaw.

Each of the two venom glands is divided into three or four sacs and each of these is independent within a fibrous sheath. Each of the sacs—as many as eight in one animal—has a central cavity or lumen which becomes narrower at one end to form a canal or narrow duct that leads to a funnel-like opening. The walls of the sacs are lined with cells filled with a granular material and it is apparently these granules that produce or at least contain the venom. No great store of venom accumulates in the central cavity of the sac. When the reptile contracts the muscular sheath around the sac, the cells secrete the venom into the cavity through countless little tubules. Once it reaches the cavity the venom is ready for the animal's use. The whole process is apparently very rapid, perhaps almost instantaneous.

A number of the heloderm's teeth are sharply grooved and by capillary action the venom fills these grooves and can be chewed into a wound. In the case of the lower teeth, the venom empties out near their base and flows upward through the grooves. The longer upper teeth fit into the funnel-shaped duct openings when the jaws are closed and can thus be soaked in venom. The venom has great spreading qualities and can be made available very quickly. In the words of Ernest R. Tinkham (1971), "On the first bite of an enraged Gila monster, approximately 35 en-

venomed teeth of both lower and upper jaws are able to introduce a lethal dose of poison into a victim; fortunately, this seldom happens." The author of those words has had personal
experience in this matter. A baby Gila monster once caught him
with a few teeth and Dr. Tinkham was hospitalized for a week.

Apparently it is only the largest of the maxillaries in the upper
jaw that reach the funnel duct openings and receive the venom.
The grooves on the leading and rear edges of the teeth are
generally flanked by sharp flanges that are really cutting edges.
The kind of wound made by such a flanged tooth would enable
the venom to spread faster and deeper than the wound made by
a simple cone-shaped tooth. Although not fangs in the sense of a
snake's hypodermic teeth, the teeth of the Helodermatidae are
distinctly specialized and adapted to the introduction of venom.
(Minton has likened them in efficiency to the fangs of the rear-
fanged snakes we will be discussing shortly.)

The system is fairly diffuse. The many ducts empty out at
some distance from some of the grooved teeth so the venom
must flow along a mucous fold between the lip and the lower jaw.
It seeps upward into the grooves in the teeth from these channels along the line of the jaw.

The venom is chiefly neurotoxic and, interestingly enough,
can be neutralized to some degree by serum from Gila monster
blood and by an extract made from the animal's liver. This indicates the presence of natural auto-antibodies. In other words, it
is true that the Gila monster is pretty much immune to its own
venom.

Present in the venom is the organic compound *serotonin*
($C_{10}H_{12}N_2O$) which not only causes an elevation in blood pressure, but also is a powerful pain producer. It may be the principal moiety responsible for the inevitable element in reports of
Gila monster bites, *pain*. There is considerable mechanical damage from a bite caused by the animal's powerful jaws and its
tenacious hold, which has been likened to that of a snapping
turtle. The symptoms, besides pain, indicate the neurotoxic pattern. Profuse sweating can be accompanied by the quick onset of
nausea and vomiting. Eventually, in repeated episodes of
nausea, the vomitus will contain flecks of blood. Thirst and a
sore throat are accompanied by dysphagia, difficulty or pain in
swallowing. Rapid breathing and general weakness approaching
paralysis have been reported a number of times. There is an
unbearable ringing in the ears (tinnitus), supersensitivity to
lights, a swollen tongue, faintness, and emotional instability. Envenomation by this animal is apparently an extremely difficult
adventure at best. Again it should be stressed that such a powerful toxin would hardly be required for food-getting by an animal
that normally steals baby mice from their nest.

The mortality rate of the Gila monster bite, if we are to go by existing records, approaches 25 percent. That would seem to be much too high. Precisely what manner of dragon is this retiring lizard equipped to defend itself against?

Bogert and Del Campo (1956) discuss 34 cases of Gila monster envenomation. Eight have been fatal according to the incomplete records available to them. Here, in chart form, derived for our purposes from Bogert and Del Campo, are the 34 cases. It should be stressed that in their description and analysis the researchers consider a number of the cases highly questionable. It should also be noted that case number 12 is the only case known at the time to involve the Mexican beaded lizard. All other cases were assumed to have involved the Gila monster.

Case Number	Captive Specimen	Victim Age	Victim Sex	Location	Year	Victim Drinking	Fatal
1	*		M	Arizona	1878	Yes	Yes
2	*		M		c1882		
3	*		M	Washington, D.C.	c1880		
4	*	55	M	Arizona	1885	Yes	Yes
5			M	Arizona	1884	?	Yes
6	*		M	Arizona	1890		
7	*		M		1891		
8			M	Arizona	1890	?	Yes
9	*		M		c1890	Yes	Yes
10			M	Arizona	c1893		Yes
11	*		M	New York	c1889	?	
12	?		M	Mexico	c1899	?	?
13	*		M	Arizona	1906		
14	*		M	New York	1907	?	
15	*		M	Germany	c1913	?	?
16	*		F	France	1911		
17	*		M	Los Angeles	1915	?	Yes
18	*	62	M	Arizona	1930	Yes	Yes
19	*		M	Arizona	1940	?	
20	?	14	M		c1930		
21	*		M		c1930	?	
22	*		M	Mexico	c1938	?	
23	*	16	M	New York	1938		
24	*	16	M	Arizona	1943		
25	*	20	M		c1943	?	
26	*	20	M	Arizona	c1943	?	
27	*		M	Arizona	1948		
28	*		M	Ohio	1948		
29	*		M	Arizona	1952		
30	*	29	M	Nevada	1953	Yes	
31	*	Teens	M	Arizona	1953		
32	*		M	Arizona	1950		
33	*		M	Arizona	1953		
34	*		M	San Francisco	1954	?	

(Since Bogert and Del Campo assembled their data, there have apparently been other cases—none, however, lethal. All involved captive animals. One, at least, involved a Mexican bearded lizard.)

The circumstances surrounding the first of these cases were reported in 1886, when a notary public by the name of Guiberson witnessed a document signed by a G. J. Hayes. In the document Hayes described the death of a miner in either 1878 or 1880, presumably as the result of a Gila monster bite. It was typical of the kind of accidents this animal has been involved in.

One day (or evening) in Tip-Top Mining Camp in Arizona a number of the boys were "whooping it up" in a saloon. There was a miner by the name of Lou Smith there as well as Hayes and a group of Italian immigrants who had come to work the mines. A miner by the name of Johnny Bostick was drunk enough to become foolish and he began taunting a 22-inch Gila monster that someone had brought in and placed on a card table. He grabbed the lizard in one hand and jabbed at it with a finger on his other. The lizard flexed its body and snapped. It caught poor Johnny's finger and amid the yowling that must have taken place

The Gila monster's only congener, the Mexican beaded lizard, *Heloderma horridum*. They are similar animals with a similar level of toxicity. Neither is aggressive toward large organisms and encounters with wild specimens are unlikely.

his friends had to pry the animal's jaws open so the probably terrified miner could get his finger free. He is said to have consumed a lot of liquor after the incident, something he clearly did not need. He became paralyzed on one side and is said to have died in three months, probably on April 19, 1878. The medical aspects of the case are difficult to determine, and it is certainly not clear what eventually killed him. Three months is rather a long time for a venom to take. He may, in fact, have simply drowned his own liver in alcohol!

In contrast to Johnny Bostick's misadventure is the story of Dr. Shufeldt of the Smithsonian Institution. He was examining a live Gila monster in the herpetological room, one that had been in captivity for nearly six months. He was holding it in his left hand and examining it with his right. Apparently the reptile was indignant at its treatment and as Dr. Shufeldt was returning it to its cage it managed to bite him on the right thumb. The bite was in the form of a severe laceration that went to the bone. The lizard did not hang on, and Shufeldt replaced it in the cage and began to work on his thumb.

He sucked on it and managed to extract some blood and probably some venom. Strangely, for a wound so deep, ". . . the bleeding soon ceased entirely." Very quickly there were shooting pains up his arms and down his side. He felt faint and there was rapid swelling of the thumb and hand. He sweated profusely. The pain was so severe he could not sleep that night, but, fortunately, the swelling did not extend beyond the wrist. Oddly enough, Dr. Shufeldt concluded from his experience that the bite of the Gila is harmless. He was, of course, quite mistaken. He probably did not receive much venom in the bite because the Gila monster did not or could not hang on. Still, the pain shooting all the way up his arm and down the corresponding side of his body should have told him something. His recovery is said to have been complete.

Of the 34 cases Bogert and Del Campo reviewed, only three, and possibly five, involved wild-living specimens. Between 29 and 31 cases involved captive specimens, which underscores the implausibility of an accident occurring in a surprise encounter as might happen with a venomous snake. All but one of the victims was male. At least five of the victims were known to have been drinking, and at least 13 might or could have been judging from the circumstances at the time of the accident. Of the eight known fatalities at least four had been drinking and three more might have been. There were two additional cases where the outcome of the encounter is not recorded—in both cases the victim may have been drinking.

What all this adds up to is that the chances are better than 30 to one that if you are bitten you will be male, the odds are about ten to one that you will be bitten by a Gila monster you are handling, there is a fairly good chance that you will be in the grip of the grape, and once upon a time there was one chance in four that you would die. All eight reported deaths, however, and the two unknown results, occurred before 1931 and there have been considerable advances in medicine generally and in the care of envenomated people specifically in the intervening four decades plus. Today it is highly unlikely that a person bitten by either a Gila monster or a Mexican beaded lizard will die unless denied medical attention.

There is an antivenin produced for the bite of the Helodermatidae and it is available without charge in the case of an emergency from the Antivenom Production Laboratory, Arizona State University at Tempe. Rabbit serum is used instead of horse. There has been some talk over the years of using cobra antivenin because it is specific to another basically neurotoxic venom. But, such ideas came in simpler days before the full complexity of venoms was clearly appreciated. Treating the victim of a Gila monster bite with cobra antivenin would be like treating a man radically for one serious disease when in fact he was suffering from another.

ENDANGERED SPECIES

Protected by law but not so often by fact in Arizona, the Helodermatidae are very likely in danger of extermination except in shrinking, isolated pockets. There is understandable prejudice against them based, of course, on ignorance of the role all predators play in nature. Man's most sacrosanct invention, the automobile, plays havoc each night as cars roar by the thousands through and adjacent to more and more of the animals' habitats. As more and more roads are built, and as more cars travel these roads, the Gila monster and eventually the Mexican beaded lizard will give way. In a strange way the fact that they are venomous makes them more often victimized by automobiles. When nature equips one of her creations with omnipotent defensive mechanisms, the creatures are never able to learn the idea of fleeing from danger. If they cannot get comfortably away without undo exertion, they hold their ground. The skunk has never quite gotten ahold of the idea that a car will be unimpressed with it. Neither has or will the Gila monster. But, skunks have ranges covering tens of thousands of times the acreage suitable to the survival of the world's only proven venomous lizards.

Chapter *9*

SNAKES

AND
SNAKEBITE

A variety of snakebite kits from the author's collection. These kits are for emergency first aid only and do not contain antivenin. They should only be used in cases where immediate professional medical help is not available.

he 2,700 species of snakes that have been described by herpetologists are highly specialized descendants of the older and much more diverse lizards. Eight to ten families are generally recognized. According to the classification provided by H. G. Dowling, formerly with the New York Zoological Society and now with the American Museum of National History, these families are:

Boidae—65 species of pythons and boas. None are venomous although many are dramatically large.*

Anilidae—11 species of so-called pipesnakes. None are venomous.*

Uropeltidae—43 species of shieldtailed snakes. None are venomous.*

Typhlopidae—at least 200 species of typical blindsnakes. All are harmless.*

Leptotyphlopidae—40 species of slender blindsnakes all of which are completely harmless.

Colubridae—a diverse collection of at least 1,400 species of harmless, or nearly harmless, snakes** plus two seriously venomous ones. The two are the boomslang (*Dispholidus*), a dangerous rear-fanged snake we will discuss in detail, and the vine snake (*Thelotornis*). Both species are African.

*Even some harmless snakes have venom glands, or produce a kind of venom in salivary glands, but for our purposes here, we will discuss only snakes with a fang system associated with venom glands.
**The Colubridae present a wide spectrum of venom and fang development. Most are relatively harmless to man. To be perfectly accurate, however, we would have to acknowledge that many Colubrids, no one knows exactly how many, are venomous to some degree.

Elapidae—230 species all of which are venomous. Included are the cobras and their allies. Specifically, the elapids are broken down into 181 terrestrial species, subfamily *Elapinae*, 4 recently evolved sea snakes, subfamily *Laticaudinae*, and 45 older sea snakes, subfamily *Hydrophinae*. (Some students place the 49 sea snakes in a separate family, *Hydrophiidae*.)

Viperidae—this family, the vipers, contains 180 species, some of which have the most advanced venomous apparatus known. Its three subfamilies are:

Causinae—23 primitive vipers (The status of this family is open to question.)

Viperinae—35 Old World vipers

Crotalinae—122 pit vipers (Some students recognize *Orotalidae* as a separate family.)

In all, this classification includes 412 species of venomous snakes, or 15.2 percent of the total number of snake species known.

SNAKEBITE AS A PUBLIC HEALTH PROBLEM

The role of the 412 species of venomous snakes in public health is a matter of concern and confusion. Many sources present startling statistics that seem to be substantiated on the one hand yet can be difficult to reconcile on the other. The figure often given for annual human mortality from snakebite is 40,000. However, most snakes in the reputedly high snakebite mortality areas have a very low mortality *rate*. The Asian cobra, a very common snake over vast areas of the earth's tropical zone, has a low mortality rate—some state it as low as 4.9 percent. For a great many snakes the percentage of people killed of those bitten is lower yet. Even if the overall figure were 9 or 10 percent, the mortality figure of 40,000 would indicate that upwards of a half million people a year are being bitten by venomous snakes.

Over 20,000 people a year are said to die from snakebite in India alone, by far the highest figure for any country. (*Not the highest rate*.) But, India does have an enormous barefoot rural population and a great many snakes. It has been suggested a number of times that the figure in India is so extraordinarily high because *snakebite* has been an expedient way to explain any inconvenient death. In villages visited only rarely by medical or paramedical teams, a person who has been disposed of in the absence of medical personnel could easily be passed off as a snakebite victim with little danger of contradiction. How true this is, how common this practice is, is really anyone's guess. We do know that snakes have been used as the instrument of homicide.

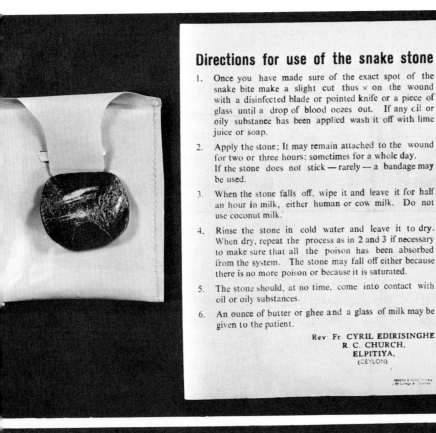

Directions for use of the snake stone

1. Once you have made sure of the exact spot of the snake bite make a slight cut thus × on the wound with a disinfected blade or pointed knife or a piece of glass until a drop of blood oozes out. If any oil or oily substance has been applied wash it off with lime juice or soap.

2. Apply the stone; It may remain attached to the wound for two or three hours: sometimes for a whole day. If the stone does not stick — rarely — a bandage may be used.

3. When the stone falls off, wipe it and leave it for half an hour in milk, either human or cow milk. Do not use coconut milk.

4. Rinse the stone in cold water and leave it to dry. When dry, repeat the process as in 2 and 3 if necessary to make sure that all the poison has been absorbed from the system. The stone may fall off either because there is no more poison or because it is saturated.

5. The stone should, at no time, come into contact with oil or oily substances.

6. An ounce of butter or ghee and a glass of milk may be given to the patient.

Rev: Fr. CYRIL EDIRISINGHE
R. C. CHURCH,
ELPITIYA,
(CEYLON)

The fabled snakestone of Asia. This example from the author's collection was obtained, as the instruction leaflet indicates, from the Roman Catholic Church at Elpitiya, Sri Lanka.

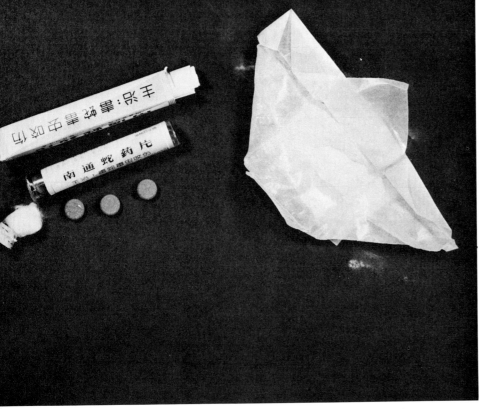

Herbal snakebite medicine from China. The tablets at left are from Peking and the directions enclosed with the vial show them being placed directly over the fang marks. The powdered material at right was obtained from an herbal doctor in Hong Kong and is mixed with lime juice and then rubbed onto wounds and taken internally. From the author's collection.

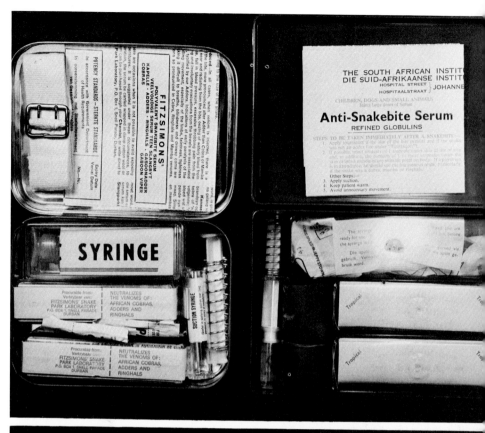

Two elaborate antivenin kits from South Africa. Both are polyvalent and cover a number of species of local snakes. The general availability of these sophisticated medicines helps to keep the human and livestock mortality rates relatively low. From the author's collection.

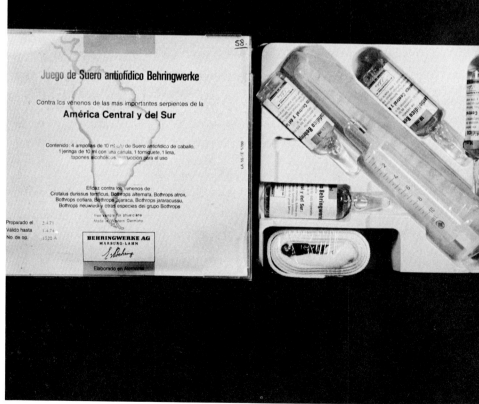

An elaborate German antivenin kit covering the more important South and Central American snakes. Kits such as this one are becoming more generally available than has been the case in the past. Not only will zoos and hospitals in large cities have them but dispensaries in small villages and towns. From the author's collection.

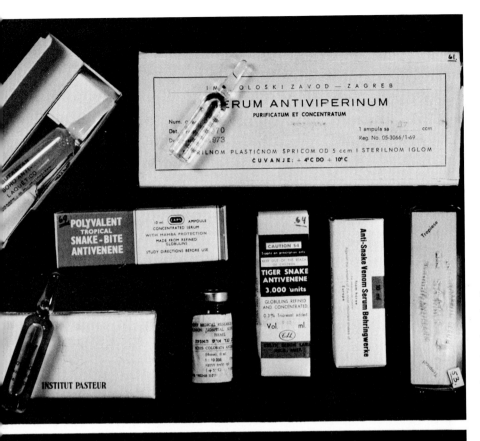

Antivenins from around the world. From the top left, clockwise: Brazil, Jugoslavia, South Africa, Germany, Australia, Israel, France, and Rhodesia. It is impossible to state what the actual reduction in human death and suffering has been since the general availability of these medicines but without doubt it has been great. Specimens from the author's collection.

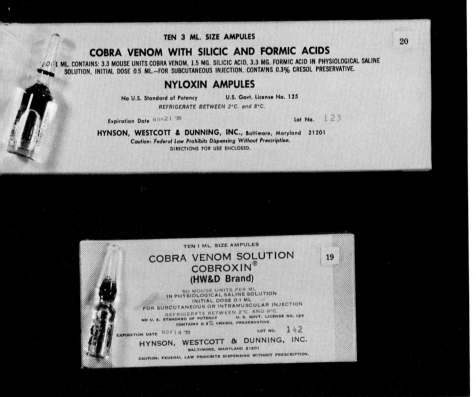

Venom as an agent of mercy. Cobra venom preparations for use in the treatment of intractable pain. From the author's collection.

(20).

AN APPEAL TO THE PUBLIC

I, the undersigned, D. H. Perera, of Temple Road, Maharagama, have the greatest pleasure in writing few words about the Moraes Visa Veda Sangamaya.

My child, Pemasiri, aged 7 years, was bitten by a viper of a dangerously venomous species (Le.polanga) on 27th September, 1958. This was a full moon-day, and the time was 9. 55 a.m. Within 45.minutes from the time of the bite, the child was rushed to the Moraes Hospital for snake-bite, at Beddagana Road, Kotte, and prompt and immediate attention was paid to the patient.

Before the lapse of 10 minutes, 4 medicines were administered, and immediate results were gained. The child was fully recovered. It is only after attending on the patient, that other particulars, such as way-fore-omen etc were inquired into. The prompt and immediate attention on this type of urgent cases is praise worthy.

All these treatments are rendered free of any charge, and even in the absence of the Chief Physician, his assistants take care of the patients in a very efficient manner.

It is needless to mention the great social service rendered by this hospital to the people of this Island. Therefore, I sincerely and earnestly appeal to the public to help to improve the one and only Hospital in the Island for Snake-bite treatment, which had saved many lives.

Yours faithfully,
Sgd: **D. H Perera**

Temple Road,
Maharagama.
5. 10. 58

ලංකාදීප විස වෙද අරමුදලින්
මොරායස් මහතාට ලබා දුන්
මෝටර් රිය.

"லங்காதீப" வாசக நேயர்களால்
திரு. மொறையாஸ் அவர்கட்கு
நன்கொடையாக வழங்கப்பட்ட கார்.

The Car donated to Mr. Moraes by the "Lankadipa" readers.

A book describing the Moraes Hospital for snakebite with a fund-raising testimonial on page 20. The small balls are the main herbal medicine used at the hospital (with a receipt for the purchase). The herbal material is meant for mixture with lime juice and is used both externally and internally.

Caras

1458

මොරායස් විස වෙද රෝහල
மொறயாஸ் விஷ்க்டி வைத்தியசாலை
MORAES HOSPITAL FOR SNAKE-BITE

ආරම්භ කළේ 13-12-57
ஸ்தாபனம்: 13-12-57
ESTABLISHED 13-12-57

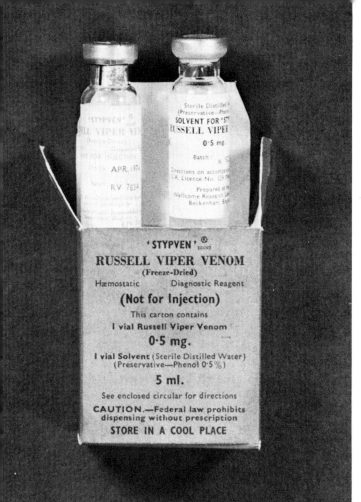

Snake venom as a diagnostic tool in the laboratory. Russell's viper venom as a reagent in blood examinations. From the author's venomological collection.

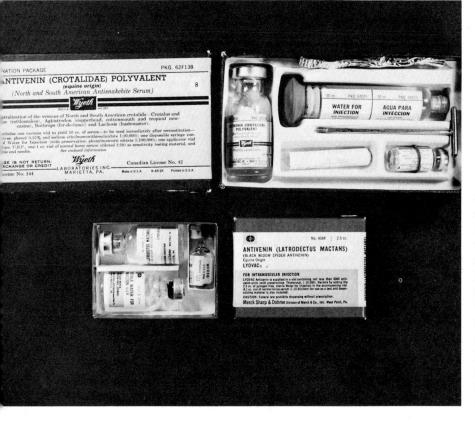

Antivenin kits from the United States. These are the only two kits available for purchase. The top kit, as indicated, covers all North American venomous snakes except the coral snakes and the bottom kit is for the black widow spider. The antivenin for the coral snake is also produced by Wyeth but cannot be purchased; it is free in the case of emergency and is stockpiled in appropriate southern hospitals. The antivenin for the Gila monster was available experimentally from University of Arizona at Tempe. Materials from the author's collection.

Monovalent antivenins obtained by the author in Bangkok. These are not kits in that they do not contain syringe, needle, or other equipment—just a 10 ml. vial of antivenin and a file for opening the sealed glass top. The Thai's do not believe in the use of polyvalent antivenins as the Europeans and Americans do.

We must remember *scale* when we discuss worldwide mortality figures. We are speaking of areas such as Asia Minor, Central America, Arizona, the Amazon basin, Kansas, South and North Africa, India, Ceylon, Burma, Thailand, Indonesia, Japan, China, Asian as well as European Russia, western Europe and the thousands of small islands in warm seas where the sea snakes abound. In plain fact, the problem of obtaining accurate mortality figures is beyond us however sophisticated our medical statistical techniques may have become. It is compounded by the fact that snakebite, lethal and sublethal, will almost inevitably be most common (we assume) in areas where there is minimum communication with medical statisticians.

Is, then, snakebite a public health problem? Decidedly so. Support for this opinion can be found in the emphasis that has been placed on research in this area and the amount of antivenin that is produced. Both of these topics will be discussed in the chapters that follow. It is simply that the *dimensions* of the problem escape us.

We should recognize one other factor in attempting to judge or even guess at the enormity of snakebite as a public health problem. For reasons best understood by psychiatrists, people almost always seem to exaggerate everything having to do with snakes. Snakes are mystical, mythical creatures that have figured in the legendry and symbology of every land they inhabit. (They are found on every continent except Antarctica.) Because of their mythical and mystical qualities they inspire the imagination to new heights of fantasy. This clear penchant for exaggeration is characterized for me in my own experience by a visit I once made to a well-known Asian snake doctor. A native rather than a true medical doctor this nonetheless educated, English-speaking gentleman had his own hospital where he treated nothing but snakebite victims. From what I was able to gather by interviewing a number of people, he was known for scores of miles in all directions. I was told of specific instances in which American and European-trained medical doctors referred snakebite patients to him. It was almost impossible to bring up the subject of snakebite in any of the villages in that part of the country without this famed man's name being mentioned. His exploits were often reported in the press.

And so I visited his hospital. It had six beds, one of which was occupied while I was there. A cobra victim seemed to be recovering nicely under the influence of this man's herbal remedies. I purchased a booklet about his work (which he had written, describing himself in less than modest terms) and some of his medicine for my collection. In answer to my questions he told me the average number of patients in the hospital at any one time was three. I had happened along on a quiet day. He further told me that the average length of time a person remained in his institution was five days. When I asked how many people he had treated his unabashed, straight-faced answer was *98,000*. By my reckoning, and his figures, this remarkable man would be in the vicinity of 450 years old. People seem to naturally say incredible things when they talk about snakes. For this reason the laboratory and the clinic have been richer sources for facts than interviews. Interestingly enough, I visited a second reknowned snakebite specialist in the same Asian country. The man, a Roman Catholic priest, sold me a snakestone. Boiled in milk (cow or human) after each use, it is supposed to be placed over a snakebite and left there until it falls away of its own accord. At that point it is said to have absorbed all the venom that was injected and to be ready to be boiled in milk and used again.

The whole subject of native remedies for envenomation is an open one and worthy of careful consideration. I have noted, for example, that many native cures in Pakistan, India, and Ceylon (and I presume other areas that I have not personally surveyed) call for lime juice. The remarkable snakebite doctor I discussed used lime juice as the base for his herbal remedy. I note an article in an English-language paper in Ceylon calling for lime juice as a first-aid item in snakebite. The theme is constant. It has been largely ignored in the West yet recent investigation has shown that viperine venoms, at least, are synergistic with ascorbic acid deficiency. That means that anyone with a low intake of ascorbic acid (in which lime juice is rich) will suffer more from viperine venom than a person with a normal level of this food substance. It is quite likely that there is a great deal more to native remedies than we have allowed for. Native doctors have had a lot of years, a lot of snakes, and a lot of snakebite cases in which and on which to work out their trial-and-error medical techniques.

There are many opinions as to the extent of the snakebite problem. In 1956 Swaroop and Grab of the Statistical Studies Section of the World Health Organization in Geneva published their paper "The Snakebite Mortality Problem in the World." Of the 412 venomous snakes fewer than 200, in their opinion, are dangerous to man. In the figures they give they acknowledge a problem of underestimation. This is inevitable, in their opinion, since statistics in the less developed countries could only be derived from cases seen by medical practitioners. In many countries only a very small percentage of the cases in any medical category will be seen at professional medical facilities. Their findings region by region, are revealing.

Africa: Snakebite mortality would seem to be of surprisingly low incidence. Although the researchers state that from the figures available ". . . it is not possible to make even an approximate estimate of the importance of the snakebite problem," they go on to *presume* snakebite deaths on that vast continent to range between 400 and 1,000 per year. When one refers to the statements by Dr. P. J. Deoras, formally of the Haffkine Institute in Bombay, in his booklet *Snakes: How to Know Them?*, ". . . *in the former Bombay State alone 1,300 persons died of snake-bite in 1955. . . .*"* one tends to see the problem as peculiarly Asiatic.

The figures that led Swaroop and Grab to their very low estimate for the great African continent were:

Country	Sample Years	Known Annual Figure or Average Number of Deaths
Egypt	1944-48	34
French Equatorial Africa	1951	83
French West Africa	1951	44
Gold Coast	1949-52	6
Kenya	1946-50	5
South Africa	1934-38	11 (European population only)

*Minton and Minton (1969) have given additional data: "In Bombay State, between 1954 and 1958 the death toll was 1,237 to 1,788 annually. . . ."

It must be stressed that these were hospital figures *only*. People bitten who died in their homes, in the care of practitioners of native medicine, and even, it is presumed, people treated in remote situations by European-trained medical personnel were not known to be included. The African figures are certainly much too low.

North America: Swaroop and Grab give the annual death rate at between ten and 20 for the United States. This, in fact, may be somewhat on the high side. The problem is negligible in Canada.

Central and South America: For many countries in this area no figures were available. Those countries included Argentina, Ecuador, El Salvador, Guatemala, Honduras, Nicaragua, Paraguay, and Peru. Statistics for the other countries are sparse.* It is known that between 1948 and 1950, 100 deaths were reported in clinics and hospitals in Bolivia from bites by *venomous animals*. What percentage was caused by snakes and what relationship this clinical report has to the country as a whole are not known.

It has been estimated that Brazil has 2,000 snakebite deaths a year. When one considers the vast areas of Brazil (a tropical nation with a wide variety of venomous snakes) which are best described as tribal, any figures must be thought of as *slightly indicative* rather than comprehensive. There are many Brazilians who never see a doctor in their lifetimes and whose fate must be completely unknown to statisticians.

British Guiana, another country with remote regions and a large number of venomous snakes, reports a very low incidence of death by venomous bites—an average of three per year between 1944 and 1948.

Venezuela reports a snakebite mortality figure rather higher than those of the other nations in the area—except Brazil. It is estimated that 150 people die each year from snakebite. How much of the differential between Venezuela and other countries of the region results from better reporting methods is not known.

It is clear that statistics from South and Central America are much too thin and the number of unknowns too great to formulate anything like an intelligent guess. The problem may in fact be much greater than we realize.

Asia: The snakebite problem is far more intense in Asia than in any other part of the world. Even the most casual statistical evidence makes this eminently clear. Swaroop and Grab estimate that Burma has the highest incidence of snakebite mortality in the world. Between 1936 and 1940 an average of 2,000 deaths were reported annually. That is a rate of 15.4 per 100,000 popu-

---

*Minton and Minton indicate 200 fatalities a year in Mexico, 25 to 35 in Costa Rica, and between 150 and 200 for Colombia.

lation. Even though India may report ten times as many deaths, the rate there is only about a third as great—5.4 per 100,000—because of the huge population differential. In some districts of Burma recorded death rates run as high as *30 or more per 100,000 of population!* The district of Sagaing reported 37 per 100,000* and Meiktila reported 34. No other political or geographical unit on earth even approaches those rates. In Burma the high snakebite incidence districts are in the lowlands, areas drained by the Irrawaddy and Chindwin rivers. The majority of deaths are apparently caused by the virulent and active Russell's viper, daboia, or tic-palonga (*Vipera russelli*)—Sherlock Holmes' "speckled band."

In Ceylon (now Sri Lanka) there were an estimated 300 deaths a year at the time Swaroop and Grab did their analysis. That figure represents only a fraction of the mortality rate found in some Burmese districts. I have visited the country three times since the study cited was done and there is no reason to believe the rate is very much lower now, nearly 20 years later. The study mistakenly omits the saw-scaled viper (*Echis carinatus*) from the list of particularly dangerous snakes found on the island. It is almost certainly a major cause of trouble in the dry zones of the North.

As indicated several times previously, India has the highest *number* (but not the highest *rate*) of snakebites each year. The annual figure is given as 20,000. This means that at least 200,000 people are bitten annually. When one considers that there are areas having only one doctor for every 70,000 people, the snakebite rate must remain conjecture. Yet there are figures like that cited from Deoras—1,300 deaths in the one state of Bombay—and these would support the statement that snakebite is a major health problem. (West Bengal has one of the highest numbers of snakebite deaths in India as well as one of the highest rates.) For the purposes of their study, Swaroop and Grab considered prepartitioned India including what are now West Pakistan and Bangladesh.

In the study sponsored by the World Health Organization, Thailand was reported to have 200 snakebite deaths each year. Considering the large number of very venomous snakes in that country, this mortality rate of 1.3 per 100,000 (very, very much less than Burma's, for example, with a comparable snake fauna) might seem small. This can in part be explained by the fact that intensive work in antivenin therapy has been going on in Thailand since 1917. The Queen Saovabha Memorial Institute in Bangkok alone treated 20,000 cases of snakebite between 1941 and 1956. As early as 1952 doctors there were treating more

*Swaroop and Grab gave the figure as 37 per *1,000* but that clearly was a printing error. The birth rate is not that high in Burma!

than 2,200 cases a year. Six species-specific antivenins are prepared at the Institute and are generally available throughout the country. From my discussions when I visited the Institute in 1967, I would be inclined to beieve it is the treatment available in Thailand for snakebite rather than the incidence of actual bites that gives the country the low mortality figure. The doctors at the Institute in Bangkok disagreed most heartily with the medical tradition in the United States and in Europe of preparing polyvalent antivenins. They likened it to treating people for diseases they did not have. They were firm in their belief that specific antivenins for each kind of snake was the answer.* Antivenins prepared in Bangkok are available at 100 hospitals and more than 2,500 health centers throughout the country.

Swaroop and Grab estimate that Asia, with the exception of China, suffers between 25,000 and 35,000 snakebite deaths a year.

Europe: Although figures for eastern Europe were not available for the World Health Organization study, Europe apparently has a low death rate from venomous snakes. Sampling figures from western Europe were given:

England and Wales—2 deaths between 1940-48
Denmark—7 deaths between 1900-1947
Sweden—15 deaths between 1915-44
Switzerland—25 deaths between 1881-1930

Europe has a small variety of vipers, generally mildly toxic snakes. Even with the enormous population density over much of Europe, the death rate from snakebite is correspondingly low.

Western Pacific: Figures for the western Pacific region are spotty, with a few exceptions. Australia, which has a number of the world's most seriously venomous snakes, reports six deaths a year on the average between 1942 and 1950—a very low rate of 0.07 per 100,000. Papua and New Guinea, also areas with a number of extremely dangerous snakes, reported a surprisingly low death rate. Between 1949 and 1952 in Papua, 118 people were seen in hospitals suffering from snakebite. Only nine died. In the same years there were 123 snakebite victims treated in hospitals in New Guinea and no fatalities.

Japan has about 100 deaths a year from snakebite according to the study. Considering the apparent casualness with which the venomous mamushi (*Agkistrodon*) are handled, this figure is not surprising.

Writing in 1962, D. C. FitzSimons (in Vivian F. M. FitzSimons' *Snakes of Southern Africa*) added some data to the Swaroop and Grab study. It was reported that China may have a greater

*Doctors elsewhere, supported by clinical evidence from many other parts of the world, are equally certain that polyvalent antivenins are the answer to snakebite mortality.

snakebite problem than was originally estimated. (Swaroop and Grab did not give figures, but suggested the number would be low.) FitzSimons says reliable reports out of Canton in 1955 gave 500 as the figure for that province alone. The article also suggests that India suffers 30,000 snakebite deaths annually and not 20,000. Although the subject of the book is South African snakes, the article specifically declines to give an estimate for that region stating that statistics are "incomplete and inconclusive." A 1955 report (Christensen: *South African Snake Venoms and Antivenoms*) again declined to make an estimate of the mortality rate in that region. However, John Visser, in his *Poisonous Snakes of Southern Africa* (1966), estimates that the snakebite deaths in South Africa and South West Africa do not exceed 10 to 14 per year.

In recent years there have been some intensive statistical analyses done of the snakebite problem in the United States. In 1966 Parrish published his summary of the figures obtained from 5,361 hospitals and 27,309 physicians. He estimated that in a sample year, 1959, 6,680 people in the United States were treated for snakebite. Of these 14 died, giving a mortality factor of 0.21 percent. Less than one person in ten million died in the United States in 1959 from snake venom intoxication. (The comparable figure in some districts in Burma would be 3,700 per ten million.) Not at all surprisingly, the rate of bites was highest in the five-to-nine-year-old group and only slightly lower in the ten-to-nineteen-year-old group. These are the ages of least discretion and greatest curiosity, and snakebite would be correspondingly high. The susceptibility of these age groups probably holds true around the world. Discretion, caution, and responsibility are age-related more than culture-related.

Interestingly, the same author (Henry M. Parrish), in a paper published in 1957, found that in an analysis of the 71 snakebite mortalities in the United States between 1950 and 1954 the fifty-to-fifty-nine age group had more than twice the number of deaths as the five-to-nine age group and exactly three times the number in the ten-to-fourteen-year-old group. Apparently snake venom is most effective towards either end of the life span of man.

Still speaking of the United States, there is an interesting statistical side note. Of the 138 fatal snakebites in the United States between 1950 and 1959, three were administered by exotic species. A night club entertainer and a snake handler were killed by cobras (*Naja sp.*) and a herpetologist was killed by a boomslang (*Dispholidus typus*). The likelihood of a physician or hospital being equipped to deal with foreign snakebites is slight, and there should be a law requiring anyone being in possession of foreign venomous snakes to also maintain under proper storage conditions an adequate supply of antivenin. Supply, in this case, should not be taken to mean a first-aid kit. Supply means enough antivenin for long-term serum therapy in the event of a really bad bite. (Current regulations would make this illegal for most private individuals, but private individuals should not keep seriously venomous snakes except in very special cases.)

It is quite clear that we have no way of really determining the full measure of the worldwide snakebite problem. Figures published as recently as 1953 for the United States, a nation known for its love of statistical data, were off by at least 200 percent. There are no figures for many areas, poor figures for others, and, as indicated, quite possibly purposefully misleading figures for yet others. The difficulty in gathering data is compounded by the fact that those areas where snakebite is likely to be most frequent and serious—the tropics and subtropics—are precisely the areas for which the figures are most unreliable.

In fact, it may never be possible to establish the true scope of the health problem resulting from snakebite. The money that would be required to do the research would be better spent on medical facilities to counteract the problem. We can state, however, without fear of rebuttal, that the snakebite problem is one that millions upon millions of people must live with and be cognizant of every day of their lives. Every year between 250,000 and 500,000 of those people will be injured by snakes—and Sherman Minton suggests the actual figure may approach one million—and tens of thousands will die as a result. Whatever the figures are they do reflect a major world health problem and that is undoubtedly why snakebite has engendered more concern and more research than most other forms of intoxication by naturally occurring toxins—that fact plus the mystical quality of the snake itself.

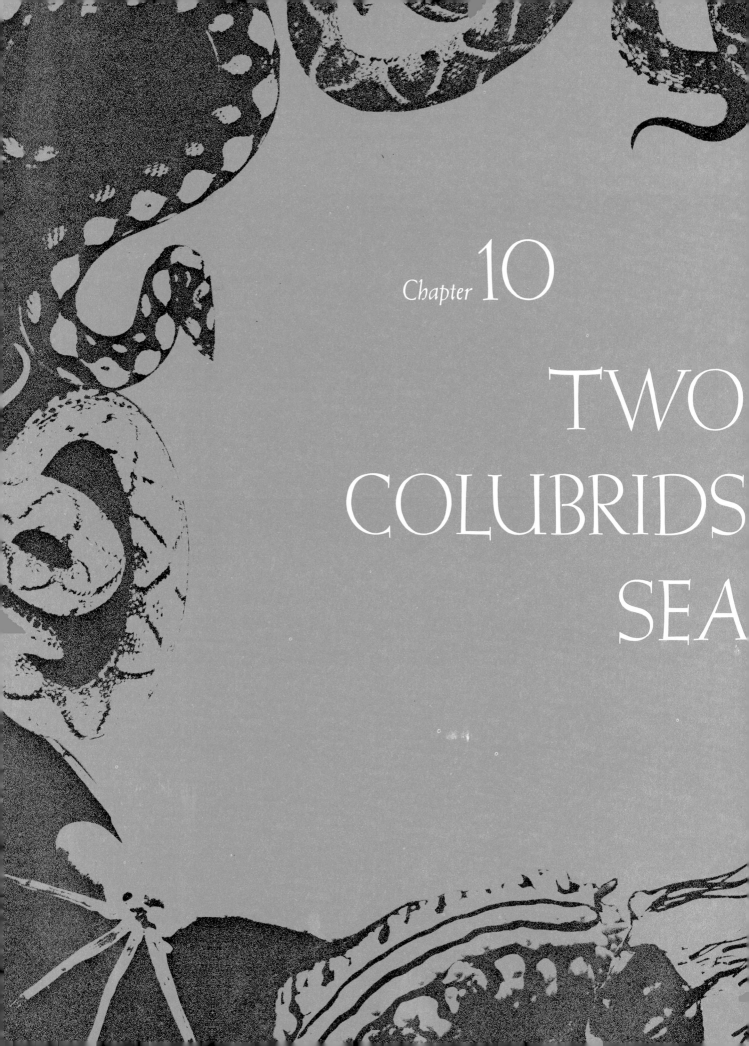

Chapter 10

TWO COLUBRIDS SEA

VENOMOUS
AND THE
SNAKES

The rear-fanged vine snake (*Thelotornis kirtlandii*), a supremely agile arboreal species from Africa. Rates as the second most venomous of the rear-fanged species. Not generally aggressive toward man and accidents are very rare with wild specimens.

One way of differentiating between venomous snakes is by fang type. Fangs are highly evolved mechanisms and are a measure of the level of evolution a snake has achieved.

The *Aglyphae* are snakes with no specialized venom-conducting teeth. Some of their teeth may be long and look superficially like fangs, but they are holding teeth and have no role in envenomation as we understand it.

The *Opisthoglypha* are the so-called rear-fanged or back-fanged snakes. They are truly venomous, but their fangs are located at the rear of the jaw and are for injecting prey after it has been caught. Envenomation apparently prevents prey animals from struggling while being swallowed and helps the snake avoid injury to delicate tissues. It is thought that neither this fang system nor the venom that goes with it evolved for striking and killing prey procurement. (Some rear-fanged snakes are also constrictors.) However, in at least two species the fangs can be used to strike, an action that would be defensive rather than food-getting in nature.

As stated in the last chapter, there are 1,400 species of snakes in the family Colubridae and all are relatively harmless to organisms as large as man except two species—both of the Opisthoglypha type. The two dangerous species are the boomslang (*Dispholidus typus*) and the vine snake, also called twig or bird snake (*Thelotornis kirtlandi*).

Both the boomslang and the vine snake are retiring, arboreal reptiles, widely distributed in the savanna areas of Africa south of the Sahara. They are not found in the rain forests of the west or in the arid region in the southwest. Both snakes are under six feet in length and are slender in build. They stick to the trees, usually, and feed on lizards, frogs, some rodents, and fledgling birds. They are typically placid and are seldom involved in an encounter with man. When they do assume a defensive posture (usually after considerable provocation), they are capable of extending their lower jaw so that it gapes and forms a flat plane with the upper jaw. In this position the fangs at the rear of the upper jaw extend outward at almost right angles and a stabbing motion is possible. This posture is usually resorted to after bluff and attempts to escape have failed.

The boomslang, *Dispholidus typus,* the most seriously venomous rear-fanged snake in the world. A retiring arboreal species from southern Africa. Encounters with wild specimens are rare and most injuries result from handling captive specimens.

The fangs of the two really dangerous Colubrids are in an intermediate phase of development. They are not the enclosed tubes of the more highly advanced venomous snakes, but they are deeply grooved and capable of conveying venom into a wound. (There are other rear-fanged snakes in South Africa that have been implicated in bites on humans—notably snakes of the genera *Trimerorhinus*, *Macrelaps*, and *Crotaphopeltis*—but encounters are of slight consequence.)

Although a bite by a boomslang or a vine snake in the wild would be a freak occurrence, there are enough snake catchers or handlers exposed to danger for study of their venoms to have taken place. Little or nothing has been done about the venom of the other rear-fanged snakes and consequently little or nothing is known about them. There is no antivenin now available for the bite of *Thelotornis*, but small quantities of *Dispholidus* antivenin are manufactured by the South African Institution for Medical Research in Johannesburg. No other antivenins work for either species.

The venoms of the seriously venomous Colubrids have a marked effect on the blood's clotting ability. In the initial stages the venom is a powerful coagulant. It is this factor, this slow-acting and powerful coagulant, that is thought to be the lethal factor. There may be moderate swelling near the bite and some discoloration. Within an hour and a half there will be severe headaches, dizziness, nausea, and vomiting. There may also be considerable abdominal discomfort. As the effects of the venom progress, the clotting reverses and extensive hemorrhaging occurs. This follows the conversion of the fibrinogen in the victim's blood to fibrin. In the first stage following this reversal, there is bleeding from the fang wounds and from any other superficial scratches. This is accompanied by internal bleeding possibly leading to collapse, coma, and even death in the most extreme cases. There may be bleeding from the mucous membranes as well. This phase is usually accompanied by a severe drop in blood pressure. There may be bleeding into the dermal and muscular tissue and livid patches may appear over the body. There will be nose bleeds and even bleeding in the eyes.

Karl P. Schmidt, the noted American herpetologist and director of the Field Museum in Chicago, died 24 hours after being bitten by a captive boomslang. He was beginning to feel better and called the museum to say that he should be expected in for work the following day. Two hours later he collapsed and died. A sixteen-year-old South African boy collecting eggs from a nest was bitten on the outside edge of his hand by a small boomslang and received treatment from the family doctor. The next day he felt better and went out in the yard, although he had been advised not to. When his parents found him he was on the ground unconscious. He was never revived.

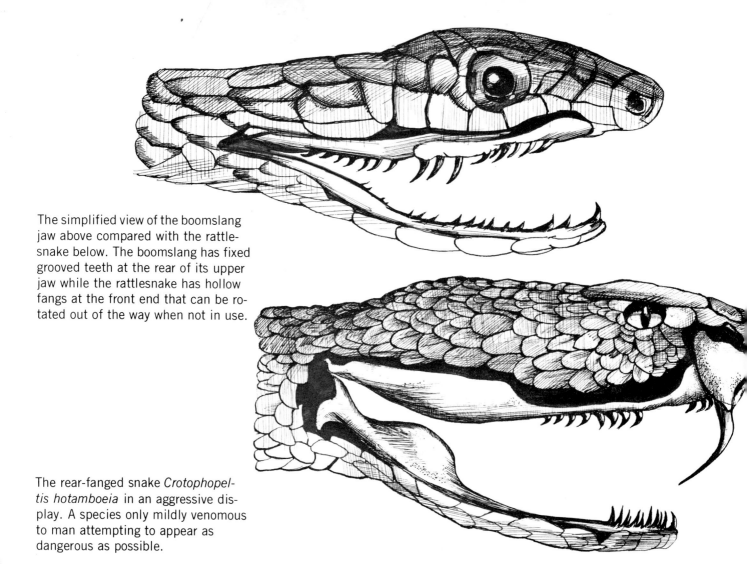

The simplified view of the boomslang jaw above compared with the rattlesnake below. The boomslang has fixed grooved teeth at the rear of its upper jaw while the rattlesnake has hollow fangs at the front end that can be rotated out of the way when not in use.

The rear-fanged snake *Crotophopeltis hotamboeia* in an aggressive display. A species only mildly venomous to man attempting to appear as dangerous as possible.

There are many cases like these, cases where victims of a boomslang bite believed a temporary easing of symptoms meant imminent recovery. They were, unfortunately, wrong. The bite of the boomslang can be a great deal worse than many people seem to realize. It is a snake to reckon with. It is most fortunate that it is not aggressive.

SEA SNAKES

As previously indicated the world's 49 sea snakes are considered by some to belong to two subfamilies of the family Elapidae—Laticaudinae (four species) and Hydrophinae (45 species) while other investigators assign them their own full family—Hydrophiidae. (For those not generally familiar with taxonomic practice, the ending *-n*ae indicates a subfamily and the ending *-d*ae indicates a full family.)

Whatever the eventual taxonomical determinations might be these marine animals are of special interest. Interestingly, there are no harmless or nonvenomous sea snakes. A well-developed venom system was apparently essential for snakes that went to sea. Snakes leave disappointing fossil records and it is not known if there ever were harmless snakes in the true salt-water environment.* Since sea snakes are thought to be related to certain elapids (notably the kraits) there probably never were harmless sea snakes ancestral to those of today.

Sea snakes are so well adapted to life in the ocean that there is a tendency to think of them as being rather eel-like and not associated with land snakes. This, of course, is not the case. Sea snakes are no less dependent than land snakes on air and are no more closely related to an eel than they are to a grouper or goldfish.

The sea snakes (we will use the common name of the group throughout since the proper taxonomic one is a matter of doubt) are of the Proteroglyphae, the more primitive of the advanced venom fang types. Their venom-injecting teeth are set far forward in the maxilla and are enclosed tubes. The principal function of the teeth may be to introduce a substance into *prey* animals that will arrest their progress or stop their struggles and make them available as food. The secondary purpose then would be defense, but this is not as clear in the case of sea snakes as with land forms. Researchers working at the Smithsonian Tropical Research Institute in Balboa, Canal Zone, studied captive sea snakes from the Pacific in tanks with potential Atlantic pred-

*There are harmless brackish water and marsh snakes: *Natrix compressicaudus*, the salt-marsh snake, is found in Florida and Cuba; *Acrochordus*, the wart snakes, are common in Southeast Asia. There are other "almost marine" snakes as well that are harmless.

ators. (Sea snakes probably have few if any predators in the Pacific since *not* eating a sea snake would be a survival advantage. Some seabirds are said to eat sea snakes.) The exercise in Balboa was to determine probable relationships between sea snakes and indigenous Atlantic animals should the sea snakes invade the Atlantic through a proposed sea-level canal. One Atlantic snapper swallowed a sea snake tail first and was bitten just below the eye before taking the final swallow. The fish died 20 minutes later. Another fish in the test tanks ate two snakes without suffering apparent harm, but died an hour later, possibly as a result of internal bites. Before dying the fish regurgitated the two snakes, both of which survived their ordeal. It was suggested that this could account for the extreme toxicity of sea snake venom. A venom that was so powerful that any animal swallowing the snake would regurgitate it quickly and unharmed after being bitten internally would be of inestimable value to a snake living in a sea surrounded by potential swallowers. There are some reports that sea snake venoms work rapidly and effectively on fish and eels which are the normal prey of these snakes.

The sea snakes represent millions of years of specialization for life under what for a reptile must be trying conditions. For one thing, like the marine iguana and the sea turtles, they had to develop a means of disposing of salt. Vertebrate body fluids have only about a third the salt concentration of sea water and, unlike marine mammals, reptiles that have returned to the sea do not have kidneys efficient enough to produce urine with heavier salt concentrations than their blood. The turtles secrete salt through a gland behind the eye, the marine iguana secretes excess salt through a gland in the nose, while the sea snakes apparently utilize a special gland under the tongue that secretes salt into the mouth. These salts are not known to be associated with the process of envenomation.

Sea snakes are found in the Pacific and Indian Oceans from the southern tip of Africa to the coast of Central and even North America in the latitudes of Baja California. The land bridge formed by Central America, separating the Atlantic from the Pacific, rose up out of the sea the last time about four million years ago. It was apparently after that time that the sea snakes reached the Western Hemisphere. If events had not happened in that order we would probably be unable to say today that there are no sea snakes in the Atlantic Ocean.

The head of the sea snake is often flattened and quite small. Those specialized for feeding on small, elongated fish would have difficulty jabbing at and biting anything as large as a man's arm or leg, but when handled they presumably all can bite. Their tails are flattened vertically and give the snakes considerable drive in the water. The advanced species like the yellow-bellied

The banded sea snake, *Laticauda colubrina*. This four-foot sea snake is common from India to the Pacific Islands. It is seriously venomous as are all sea snakes.

The yellow-bellied sea snake (*Pelamis platurus*) is the most wide-spread venomous snake in the world—north of Japan to South America. Accidents are surprisingly few in number but very serious when they do occur.

sea snake (*Pelamis platurus*) are quite helpless on land and can do little more than flop around. When driven up onto a beach by onshore currents they are usually doomed to a fairly quick death. They have a very low tolerance for heat and probably perish with exposure to 91 degrees fahrenheit for any length of time. Unless they can flop back into the water and submerge, they are doomed on a sun-drenched stretch of sand.

Our ignorance of the habits and ecology of the sea snakes is profound. We do not seem to have the answers to even the most basic questions about these animals. As an example, in the Strait of Malacca, the passage between Sumatra and the Malay Penin-sula, there are believed to be 27 forms of sea snake. How do they not compete?

The slow metabolism of the sea snakes allows them to remain submerged for considerable lengths of time. Periods ranging from two to eight hours have been reported and may be accurate. The lungs of these animals are greatly enlarged and in some species the single lung extends all the way down the body to the base of the tail. That may be to facilitate long submergence and deep dives and, possibly, also to achieve trim. How the sea snake achieves pressure equilibrium in deep dives—perhaps as deep as 100 fathoms in some species—is not known. (The sea snake *lung* is as much an air sac as a functional lung.)

The less specialized sea snakes (*Laticauda*) are egg-layers and therefore shore breeders. They lay their eggs in caves and crevices and leave them to fate. The advanced sea snakes never leave the water and their young are born alive in the sea.

It is thus in the context of a highly specialized way of life that we must view the envenomating system of the seasnakes. They are animals remote from the experience of most of the Western world, but for fishermen and other people who live close to the sea in Asia and East Africa they are an important consideration in each day's activities. Many forms are eaten and one fishery in the Philippines takes the skins of about 100,000 sea snakes a year. There are more than a few records of human fatalities where these animals are regularly handled.

The relatively uncommon Stoke's sea snake, *Astrotia stokesii*. Little is known about this species and there have been reports of enormous concentrations of them in the spring, particularly in Malacca Strait. It grows to be five feet long.

All of the sea snakes have fixed, rather small, and fragile fangs at the forward end of the upper jaw. The fangs are generally well hidden by a fold in the mucous membrane of the gums and can be easily missed in a cursory examination. However, in all sea snakes they are present and form enclosed hypodermic tubes. They are attached by large calibre ducts to venom glands on either side of the head.

Because of their small heads and very small fangs the sea snakes do not strike or stab the way so many land snakes do. They grasp and hold and chew, this being true whether the victim is a fish or a fisherman. It has been reported that no more than 25 percent of the sea snake bites cause envenomation. (It must be understood that there is a distinct difference between the bite of a venomous snake and envenomation. Invariably there are some "dry" bites.) The low percentage of venomous bites may represent a reluctance on the part of the snake to release the venom it needs to obtain food or it may mean a snake has recently fed and exhausted its venom supply. In general, sea snakes discharge less venom than comparably sized land species. They do not need as much since sea snake venom is usually between two and ten times as lethal as the venom of the most dangerous terrestrial species.

It has not been determined how sea snake venom works. Tests on laboratory animals almost invariably produce paralysis, which is also reported for human victims. However, lesions seen in the skeletal muscles of human bite cases apparently cannot be obtained by injecting experimental animals with venom. Much research remains to be done.

The local symptoms of a sea snake bite are few or may even be entirely absent. If the bite occurs in murky water, the snake may retire without the victim's even realizing he has been bitten. There is no swelling, no discoloration, and no pain beyond a pin-prick sensation. There may be a gradual loss of sensation with numbness spreading to areas adjacent to the bite. There can be a slight sensation of ants crawling over the skin (formication.)

There is usually a quiet period after the bite that runs from half an hour to an hour and a half. (This period may be as brief as five minutes or as long as eight hours—both situations have been reported in clinical notes.) A relationship between the length of the quiet period and the quantity of venom injected has not been clearly established. There then ensues a depression of tendon reflexes that later will disappear. It becomes more and more difficult for the victim to move his limbs. Paralysis sets in. Death, when it follows, usually comes as a result of respiratory failure. In laboratory animals the heart often continues to beat long after the animal has stopped breathing.

Other symptoms include severe muscular pain, red-colored urine, and *trismus*, or difficulty in opening the mouth. This latter symptom is a result of venom injury to the straited muscles of the jaw. It is likely that the eyelids will become paralyzed early in the onset of the syndrome, a reaction that.has given rise to the mistaken belief that victims fall asleep and then die. In fact, from clinical observations, those human victims who do succumb are conscious up until the time of death. It is believed now that the mortality rate is less than 17 percent. It was once thought to be considerably higher.

There is general agreement that sea snake venoms, some of them at least, are the most toxic snake venoms known. Fortunately, although enormous concentrations of sea snakes are reported at certain times each year (these vast assemblages may be due to feeding patterns near estuaries during the rainy season, or they may have something to do with reproductive behavior), and although the areas where sea snakes are most commonly found are in the seas adjacent to some of the most heavily populated land masses in the world (and areas heavily dependent on the sea for food), serious encounters are not very common. Compared to land snakes the sea snakes cause a relatively small part of annual snakebite mortality among humans.

Sea snakes may be especially dangerous at certain periods of the year and indeed are said to be more aggressive when in an apparent breeding pattern, as might be expected. Some sea snakes may be only marginally dangerous to an organism as large as a man, but that is not known. All sea snakes should be considered dangerous at all times and treated with profound respect. An antivenin specific to the very dangerous beaked sea snake (*Enhydrina schistosa*) is manufactured by the Commonwealth Serum Laboratory in Melbourne. (There have been varying reports as to aggressiveness from different parts of the sea snakes' range. If these animals are migratory, such differences in behavior could be accounted for by the different activities engaged in at different ends of the migration route; patterns, as suggested, such as breeding.)

In his 1956 paper H. A. Reid gave a report of sea snake bite cases in Malaya. In 1954, an eight-year-old Chinese boy was wading in shallow water ten feet from shore. He looked down and saw a snake fastened to his ankle. There was no pain, but he was frightened and showed the bite marks to his father who was sitting on the beach. The father let the boy go back into the water to play. Within an hour the child began to feel stiffness and muscle weakness. Four hours after the bite he was drowsy. At midnight there was blood in his urine and at that point his father, who appears to have been somewhat relaxed about the whole thing, decided to call a doctor. Four hours later the boy's breathing became rapid and rather noisy and the doctor decided to send him to a hospital. At that point the doctor suspected poliomyelitis, which suggests that the father had not bothered to tell him about the snake.

The boy's condition continued to deteriorate and by 6:00 A.M., 13 hours after the bite, he was dead. One cannot help but feel there was a degree of culpability on the part of the father as well as the snake in this case.

On the same beach, two hours before the eight-year-old boy was bitten, a twenty-six-year-old man was swimming in water chest deep. He felt a pin-prick sensation on his foot but could not see anything when he examined it. Within an hour he was feeling stiffness and pain. Three hours later he had difficulty swallowing. He was admitted to the hospital 17 hours after the encounter. Despite all that was done for him he died three days later.

This long period before death is not unusual, and in the third case presented by Reid an even longer period of time elapsed between bite and death. A fifty-four-year-old fisherman was admitted on November 24, 1952, *eight days* after being bitten by a sea snake but, he did not die until November 28, 12 days after the bite.

The sea snakes, at least *some* sea snakes, are by far the most widely distributed snakes in the world. More kinds are found in the Indo-Australian area than anywhere else and, indeed, it is thought that the snake originally returned to the sea in this area. If there are fossil remains of ancestral forms this is where they will be found in all likelihood. Halstead (1970) gives the following listing of the sea snakes of the world and their ranges:

SPECIES	COMMON NAME	DISTRIBUTION
Acalyptophis peroni	Peron's sea snake	Hong Kong, tropical Australia
Astrotia stokesi	Stoke's sea snake	Makran coast, Ceylon Singapore, north coast of Australia
**Enhydrina schistosa	Beaked sea snake	North coast of Australia, Vietnam to Persian Gulf
Hydrelaps darwiniensis	Port Darwin sea snake	North coast Australia, south coast New Guinea

(Species implicated in human envenomations are marked with an*. Known human *fatality* implication is indicated by **.)

Species	Common Name	Distribution
Hydrophis belcheri	Small-headed sea snake	Gilbert Islands, Fiji, Celebes, Philippines
Hydrophis bituberculatus	sea snake	Colombo, Ceylon (Sri Lanka)
Hydrophis brooki	Brook's sea snake	Strait of Malacca, north to Trang, Gulf of Thailand, Vietnam, to Borneo, Java
Hydrophis caerulescens	Banded sea snake	China, Malay Archipelago, to Bombay, India
***Hydrophis cyanocinctus*	Annulated sea snake	Japan, Malay Archipelago
Hydrophis elegans	Elegant sea snake	Tropical Australia
Hydrophis fasciatus	Banded small-headed sea snake	Strait of Malacca, Burma, India
Hydrophis inornatus	Malabasahan sea snake	Australia, Philippines, Java, China
Hydrophis kingi	King's sea snake	Tropical Australia
**Hydrophis klossi	sea snake	Gulf of Thailand, Strait of Malacca
Hydrophis lapemoides	sea snake	Coasts of India and Ceylon to Persian Gulf
* *Hydrophis major*	sea snake	Subtropical and tropical Australia
Hydrophis mamillaris	sea snake	India
Hydrophis melanocephalus	Black-headed sea snake	Ryukyu Islands, Taiwan
Hydrophis melanosoma	sea snake	Borneo, Makassar, Strait of Malacca
Hydrophis mertoni	Merton's sea snake	Aru Islands
***Hydrophis nigrocinctus*	sea snake	Bay of Bengal, Burma, possibly Malay Archipelago
***Hydrophis obscurus*	Banded sea snake	East coast of India, Mergui Archipelago
***Hydrophis ornatus*	Reef sea snake	New Guinea, China to Persian Gulf

Venomous Animals
of the World

Species	Common Name	Distribution
Hydrophis semperi	Semper's sea snake	Lake Bombon, Luzon (only freshwater species)
**Hydrophis spiralis*	Yellow sea snake	Malay Archipelago to Persian Gulf
Hydrophis stricticollis	sea snake	Burma south to Gulf of Martaban, India north of Orissa
Hydrophis torquatus	sea snake	Strait of Malacca
**Kerilia jerdoni*	Jerdon's sea snake	Strait of Malacca, east coast of India, Ceylon, Mergui Archipelago
Kolpophis annandalei	Annandale's sea snake	Vietnam, Malay Peninsula, Java
Lapemis curtus	sea snake	East coast of India to Saudi Arabia, southern Japan
**Lapemis hardwicki*	Hardwicke's sea snake	Northern Australia, to Mergui Archipelago
Microcephalophis cantoris	sea snake	Malay Peninsula to Burma and India
**Microcephalophis gracilis*	Graceful sea snake	Southern China, northern Australia, to the Persian Gulf
Pelamis platurus	Yellow-bellied sea snake	Indo-Pacific area, southern Siberia to Tasmania, western Central America to eastern Africa
**Thalassophina viperina*	sea snake	Southern China, Malay Archipelago to Persian Gulf
**Thalassophis anomalus*	sea snake	Moluccas, Borneo, Java, eastern Sumatra, Gulf of Thailand
Aipysurus apraefrontalis	sea snake	New Guinea, Philippines, Borneo
Aipysurus duboisi	DuBois' sea snake	New Guinea, northern Australia, Loyalty Islands
Aipysurus eydouxi	Spine-tailed sea snake	Vietnam, Thailand, Indonesia, Queensland
Aipysurus foliosquama	sea snake	Ashmore Reefs, Timor Sea

Species	Common Name	Distribution
Aipysurus fuscus	sea snake	Celebes, Timor Sea
Aipysurus laevis	Olive-brown sea snake	New Guinea, tropical Australia
Aipysurus tenuis	sea snake	Northwest Australia
Emydocephalus annulatus	sea snake	Tropical Australia, Loyalty Islands
Emydocephalus ijimae	Ijima sea snake	Ryukyu Islands, Southern Taiwan
Laticauda colubrina	Yellow-lipped sea snake	Widely distributed: tropical Indo-Pacific islands, southern Japan to Bay of Bengal
Laticauda crockeri	Crocker's sea snake	Lake Tungano, Rennell Island, Solomon Islands
Laticauda laticaudata	Black-banded sea snake	widely distributed: tropical Indo-Pacific islands, southern Japan to Bay of Bengal
Laticauda schistorhynchus	Samoan sea snake	Fiji, Samoa, Tonga, Niue Islands
Laticauda semifasciata	Erabu-Unagi sea snake	Philippines, Moluccas, Ryukyu Islands

Taxonomists do not agree on some of the species listed above. Minton, who read this chapter, does not agree with the species validity of *Laticauda schistorhynchus, Emydocephalus ijimae,* or *Hydrophis melanocephalus.* Other references reviewed differed with Minton on these and with each other on several additional species. There is no way, really, of making all taxonomists happy. Since our subject is not taxonomy, or even herpetology, some way around this dilemma must be found. I use the very simple expedient of selecting a reliable source and sticking with it. The listing of sea snakes given above is according to Halstead. It is possible that even he would make certain changes in a list as long as this, but there is no way of keeping taxonomic lists up-to-date except by using a mimeograph machine and issuing weekly bulletins! What I have said here about the sea snakes will apply in the other chapters on snakes that follow—except, of course, Dr. Halstead will not be the source.

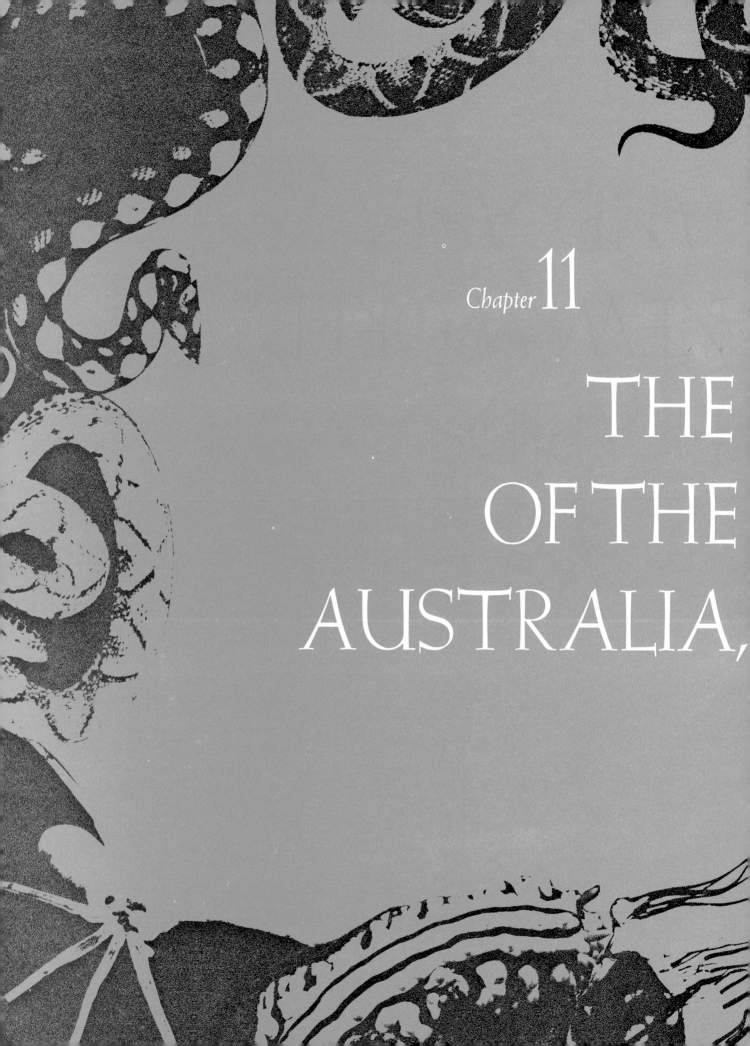

Chapter 11

THE
OF THE
AUSTRALIA,

ELAPIDS NEW WORLD, AND ASIA

The elapid fang system with fang
length exaggerated. Fixed, immobile
fangs are located at the front of the
upper jaw and must be accommodated
in position when the snake's jaws are
closed.

he cobras and their allies belong to the family Elapidae. In this group are 181 terrestrial snakes including some of the most aggressive and most seriously venomous in the world. Their fangs are of the Proteroglyph type, relatively small enclosed tubes, fixed and immovable, situated in the front of the upper jaw. Some few are especially adapted for spitting venom as well as injecting it. The range of toxicity among the elapid snakes is enormous, from deadly to almost harmless.

CORAL SNAKES IN THE AMERICAS

With the exception of one group the Elapidae are old-world snakes, found in profusion in Africa, Asia, and Australia. The only representatives in the Western Hemisphere are the coral snakes.

Area	Number of Species	Genera
North America (United States only)	2	*Micrurus* *Micruroides*
Central America and Mexico	20	Micrurus
Tropical South America	34	*Leptomicrurus* *Micrurus*
Temperate South America (Argentina and Uruguay)	4	*Micrurus*

This listing indicates that in all of the Western Hemisphere only 60 of the 181 elapid species occur, only three out of the 41 genera. It also illustrates very clearly the universal truth that reptiles are more diverse in the tropics and thin out as the latitude goes up, North and South.

The three genera that are found in the New World are exclusive to it. Because the snakes representing these genera are so different from the other venomous snakes of the hemisphere, they have been the subject of a number of myths (as, indeed, have all snakes). It is believed by a great many people, for example, that the bite of a coral snake cannot be survived by man. This is, of course, pure nonsense. Between 1962 and 1967 there were 33 coral snake bite cases in Florida with *no* fatalities. (Death did occur in one additional case.) In the same period 14 people were killed by other snakes in that state.

A second myth about the coral snake is the supposed rapidity with which its victims succumb. It is variously reported that anywhere from ten seconds to one hour is all the time a poor wretch has left on earth once he has been bitten. Of course, inevitably, there are stories that death from the coral snake bite comes automatically at sunset. This kind of nonsense is not unusual nor is it unexpected. Antarctica is the only continent not reputed to have a reptile whose bite is either inevitably fatal or instantly fatal, or both. Only the fact that Antarctica has no snakes has spared it this type of foolishness. Parallel beliefs are not difficult to find. A krait (another elapid) in India is known by a native name the translation of which is "seven stepper." The expressed implication is that the victim has time for exactly seven steps before dying. There is a "hundred stepper," too. Strange that a snake whose venom is so mild the victim can actually take a hundred steps before dying (inevitably, of course) should even warrant recognition!

The two to four foot long coral snakes (all are small animals) closely resemble species found in Asia, and it is automatically suggested that a land bridge must have once existed and that an invasion took place. All of the New World elapids resemble each other closely enough to allow for divergence from a single invading species.

The distinguishing characteristic of a coral snake bite in the context of the Western Hemisphere is the fact that it is generally quite painless. Most of the other venomous snakes of the New World have essentially hemotoxic venoms and produce a bite that ranges from painful to agonizing. The bite of the coral snake, in which a neurotoxic substance is introduced, is usually quite different.

The fangs of the coral snake are extremely small (a magnifying glass may be needed to locate them in the mucous folds at the front of the maxilla). Since the coral snake is a small animal with a correspondingly small head and mouth, it cannot strike the way larger snakes do. It must obtain purchase and chew, literally that, for as long as a minute. McCollough and Gennaro (1963) present the following coral snake bite case representing many aspects of typical coral snake envenomation.

In July, 1963, a husky, six foot, thirteen-year-old boy was bitten by a coral snake. He was cutting grass near his family's fishing camp in Florida when he saw the snake trying to get away and caught it. The time was 9:15 A.M. He did not know the identity of his catch until a companion advised him of its dangerous nature. The youngster apparently panicked and lost his hold on the snake's neck. It managed to turn and bite him on the end of his finger. The snake held on and chewed for a period of time that may have exceeded one minute. The frightened boy tried to shake and then pull the snake loose, but to no avail. It was finally killed and was subsequently identified by a herpetologist as a North America coral snake *Micrurus fulvius fulvius* 22.5 inches long. It was not until an hour and a quarter after the accident that the boy was first seen at the hospital. He was badly frightened but showed no symptoms of illness. There was no pain, no swelling until incisions were made and suction applied. He was walking and talking normally. His blood pressure and temperature were within normal limits. A complete physical examination revealed no abnormalities.

As soon as the snake was identified antivenin prepared in Brazil specific to another coral snake of the same genus, *M. corallinus*, was ordered. It was received at the hospital within 45 minutes. The antivenin was administered along with antibiotics, tetanus toxoid, and other supportive therapy. Urine and blood studies were normal and suction over the wound was continued for two hours. A tourniquet was applied and released at intervals.

The boy was placed in a room with a private nurse. Still, the only sign that he had been injured was the swelling and soreness at the site of the bite, caused by the treatment rather than the bite itself. Then, at 5:00 P.M., approximately seven and a half hours after his encounter with the snake and five and a half hours after the beginning of antivenin therapy, he complained of difficulty in breathing and was seen to be salivating profusely. He did manage to swallow some liquids. Paralysis was first noted in the muscles controlling his eyes and then it began to progress. He lost the power of speech. The reflexes in both legs were recorded as hyperactive. He responded to questions by wiggling his toes and elevating his scalp. He stated that he felt numb, but had no pain. At 7:17 P.M., ten hours after being bitten, he was placed in an iron lung to compensate for irregular and shallow breathing. Antivenin was again administered. Mucous collected in his throat and he had to be relieved of it by suction. A plastic airway was inserted. At 11:00 P.M. a retention catheter was inserted.

By 5:30 the following morning the boy was unresponsive except that he could still raise his forehead. Otherwise he was totally paralyzed. By 6:30 A.M. his pulse had become irregular. At 9:30 his heart abruptly ceased to beat. He had lived for 24 hours and 15 minutes after being bitten on the tip of his finger by a snake less than two feet long.

The Arizona coral snake *Micruroides euryxanthus*, a small and inoffensive snake rarely involved in accidents with man. It is seriously venomous, however, and not an animal to be handled carelessly, if, indeed, it must be handled at all.

The unfortunate boy in Florida succumbed to the elapid neurotoxic factor, one that typically causes paralysis. With some modifications (but no more than could be found between bites by the same species on different victims) the boy might have been bitten by a snake in Africa, Asia, or Australia. Part of the mysticism that surrounds these particular snakes may come from the fact that a bite that is capable of killing a man will cause him no immediate pain, and perhaps no pain at all. If something hurts and hurts badly, evil consequences are not so mystifying. It is the "sneakiness" of elapid venom that may be even more disconcerting for the victims and those who witness their travail, the way a neurotoxin "sneaks up on you."

The claim that the venoms of the various coral snakes exceed all others in virulence is simply not true. They are much more potent than some, less potent than others. Referring to Russell and Puffer (1971), it is possible to draw a rough scale against which to consider this matter. From the results of one series of tests, introducing the venom of 25 kinds of snakes into laboratory animals intraperitoneally, we can surmise the following for the venom of the Eastern coral snake (*Micrurus fulvius*):

Its venom is:

10.82 times as potent as that of the copperhead* (*Agkistrodon contortrix*)

5.26 times as potent as that of the cottonmouth (*A. piscivorus*)

6.11 times as potent as that of the bushmaster (*Lachesis mutus*)

1.94 times as potent as that of the Eastern diamondback rattlesnake (*Crotalus adamanteus*)

3.82 times as potent as that of the Western diamondback rattlesnake (*C. atrox*)

However, coral snake venom proved to be:

¼ as virulent as that of the Mojave rattlesnake (*C. scutulatus*)

1/3 as virulent as that of the tropical rattlesnake (*C. durrissus terrificus*)

½ as virulent as that of the Indian cobra (*Naja naja*)

1/24 as virulent as the venom of the Australian tiger snake (*Notechis Scutatus*)

*In another series of tests (Vick *et al.*, 1967) generally higher readings were obtained by injecting mice intravenously. Toxicity of coral snakes over other species ran: Southern copperhead 15.35/1, Northern copperhead 11.59/1, cottonmouth 8.48/1, Eastern diamondback rattlesnake 4.01/1, and Western diamondback rattlesnake 1.87/1.

Elapid venoms affect neuromuscular transmission apparently without having much of an effect on the central nervous system. They kill by causing respiratory paralysis. No venom is simple, far from it. There may be several neurotoxic factors in a single venom and even the term *neurotoxin* can be confusing because it is so often used without discrimination. It is a kind of catch-all that is thrown around as if it should immediately make everything clear. It does not. (Any more than the word *instinct* does.) It is an ill-defined word and a neurotoxic venom may contain hemotoxic factors and cardiotoxins as well. Actual neurotoxins are fractions of whole venoms. They can be extracted and can be decidedly lethal. In this book the word neurotoxin should be taken to mean a venom that is *essentially* neurotoxic in its effect.

There are some antivenins available for the New World elapids, but they do not begin to cover the field. The Instituto Butantan in Sao Paulo, Brazil, produces an *anti-elapidico* for *Micrurus* and it is said to be effective for an indefinite number of coral snake species. In clinical situations it can have been tried on envenomations by only a limited number of species. Wyeth Laboratories in Phildelphia prepare an antivenin for the Eastern coral snake (*M. fulvius*). It is not effective against the bite of the Arizona coral snake (*Micruroides*), a mildly toxic species. The antivenin is not sold but is stored in the nine states where the Eastern coral snake occurs. It is available without charge to physicians treating bona fide coral snake envenomations.

ELAPIDS OF AUSTRALIA AND NEW GUINEA

Following no particular geographical logic but attempting, rather, to take the large and formidable Elapidae in manageable units, we will discuss those species found in Australia and New Guinea next.

As will have become apparent, Australia has more than its fair share of venomous animals. From the funnel web spiders to the blue-ringed octopus and stonefish, from the sea wasp to its herpeto-fauna, Australia is a strange epicenter for naturally occurring lethal toxins. The Elapidae make a very real contribution to Australia's outstanding record.

A great many Australians have a blind terror of all snakes. Many believe (and some very educated people have told me) that there are no harmless snakes in Australia. (There are.) They further believe, not at all surprisingly, that there are snakes whose bite cannot be survived (such an animal does not exist) and that Australian snakes, unlike the snakes of other lands, seek out human beings to attack. While it is true that there are some particularly virulent and even aggressive snakes in Australia, none of these beliefs is true. It could be pointed out that Australia has enough trouble with the snakes it does have without adding mythical dimensions to the problem.

ELAPIDAE OF THE AUSTRALIAN REGION   (*Indicates genera dangerous to man.)

Genus	Number of Species New Guinea/Solomons	Number of Species Australia/Tasmania
*Acanthophis death adders	1	2
Apistocalamus burrowing snakes	5	0
Aspidomorphus crowned snakes	2	6
Brachyaspis desert snakes	0	5

Brachyurophis
  girdled snakes
  Australian coral snakes
	0	3

*Demansia brown and whip snakes	1	17

*Denisonia
  copperhead
  ornamental snakes
	1	6

Elapognathus little brown snake	0	1
Glyphodon collared snakes	1	3
Hoplocephalus broad-headed snakes	0	3

*Micropechis
  Pacific coral snakes
  small-eyed snakes
	2	0

*Notechis Australian tiger snake	0	1
*Oxyuranus Taipan	1	1
Parademansia	0	1
Parapistocalamus Hediger's snake	1	0

*Pseudechis
  black snakes
  mulgas
  king brown snake
	2	3

Pseudopistocalamus Nyman's snake	1	0
Rhinocephalus Muller's snake	0	1
Rhynchoelaps desert banded snakes	0	2
Toxicalamus elongate snake	2	0
*Tropidechis rough-scaled snakes	0	1
Ultrocalamus short-fanged snake	1	0
Vermicella bandy-bandys	0	5

All of the venomous land snakes of Australia, New Guinea, and the Solomon Islands belong to the Elapidae. The Viperidae and Crotalidae are not represented by so much as a single species. The only remotely comparable situation is to be found in western Europe where the only venomous snakes are Viperidae. The parallel is a shaky one, however. There are seven species of venomous snakes in western Europe while, between them, Australia, New Guinea, and the Solomons have an estimated 82.

From the preceding list it should be easy to measure the validity of the outlandish claims made for Australian snake fauna. Of the 17 genera of elapids only seven have been adjudged dangerous to man. No doubt some of the others could envenomate an organism as large as an adult human being if improperly handled, but they certainly do not justify alarm, much less hysteria. Interestingly enough, Australia also has six rear-fanged colubrids that are harmless to larger animals. They do not approach the South African boomslang in toxicity.

The Commonwealth Serum Laboratories have been studying the venoms of Australian fauna with an intensity and to a degree equaled in few other lands. They list four principal toxic elements in the venoms of Australian snakes. The relative quantities of these toxic factors will determine the syndrome experienced by a bite victim. The four factors are:

(1) *Neurotoxins:* These, as we have seen, are commonly the predominant factors in elapid venoms. They apparently block or seriously impair communication between nerves and muscles. They do not appear to attack the central nervous system. This group of lethal or impairing fractions (assuming more than one neurotoxic factor) is the most common group in the venoms of the dangerous Australian elapids.

(2) *Haemorrhagins:* This is a highly destructive group of enzymes which break down tissue. Blood vessels and therefore internal organs with a rich blood supply suffer from these enzymes. No doubt this group has evolved from digestive enzymes. In fact, haemorrhagins injected into a snakebite victim begin a process closely resembling digestion.

(3) *Haemolysins:* As its name implies (hem- or haem- from the Greek meaning blood and ly- from the Greek meaning to loosen, dissolve, or break up) this group of substances acts to break up red blood cells. They are especially notable in the venoms of the genera *Pseudechis* and *Denisonia*. They are not the principal lethal factors in Australian elapid venoms.

(4) *Thrombase:* This factor is a coagulant. (From the Greek *thromb-* meaning clot.) In a major blood vessel this factor alone could prove lethal.

The major factors were evolved for food-getting. We must keep that in mind as we review the Australian snakes. These factors, singly or in combination, are for halting the flight of, stopping the struggles of, and enhancing the digestibility of prey animals. For no species is man the prey. It happens, unfortunately for man and snake alike, that these substances have a secondary use, defense. And defense is the only thing that can prompt a snake to strike anything as formidable as a man, in Australia or anywhere else.

THE TAIPAN AND OTHER AUSTRALIANS

By anyone's reckoning the taipan (*Oxyuranus scutellatus*) is potentially one of the world's most dangerous animals. There is a form in Australia and another in coastal New Guinea. It is one of the largest venomous snakes in the world (specimens up to 11 feet have been reported) and delivers one of the most massive doses of venom. Not only is the dosage large, but drop for drop it is one of the most deadly in the world. Essentially neurotoxic it is believed to be at least twice as dangerous as the venom of the terribly and justifiably feared tiger snake.

Like most snakes the taipan is retiring, apparently preferring (no conscious preference implied but, rather, a survival-oriented behavioral characteristic) to retain its venom for the important business of food-getting. When aroused, however, apparently when feeling itself threatened, the taipan can stage a particularly alarming display. It flattens its head and neck vertically and raises part of its forebody off the ground. The tail is also elevated and waved back and forth. The snake's strike is swift and frequently unexpected. It is so rapid that victims are often struck several times before being able to withdraw to a safe distance. The bulk of the venom administered may be given in the first bite. It is not clear why a snake delivering a massive dose of so highly toxic a venom should be programmed to strike more than once thereby increasing whatever risk exists for its own safety in a combat situation.

The recovery rate from the bite of the taipan is low—quite possibly the lowest for all snakes. It has been given many times and many different figures have been used, however the taipan is less often incriminated in bite cases than several other species and figures may be unreliable. The symptoms of intoxication are vomiting, flaccid paralysis, bulbar and eventually respiratory paralysis. There may be peripheral circulatory failure. Blood involvement is slight. Until the availability of a specific antivenin, the prognosis for a taipan bite victim was poor. It was guardedly stated that "recovery was unlikely." The chances for a victim are now very much better. The antivenin specific to the taipan (and also useful, apparently, in treating the bites of several other Australian elapids and possibly also some Asian cobras and kraits—this latter point requiring further clarification) is manufactured by the Commonwealth Serum Laboratories in Melbourne.

Hardly less dangerous than the taipan, and responsible for more bites each year, is the Australian tiger snake (*Notechis scutatus*). It is the most dangerous snake in southern Australia and has a very powerful venom. Specimens may exceed six feet, notably in Victoria and Tasmania. There is a record length of eight feet on Chappell Island. The tiger snake has one of the most potent venoms known and is the most heavily studied of all Australian snakes. Symptoms in a human bite victim include pain, vomiting, a general dulling of sensation, and a kind of drunkeness that progresses until it resembles brain damage. The victim displays an inability to express ideas, and has difficulty in swallowing. There is profuse sweating, respiratory paralysis, and circulatory collapse. A specific antivenin is prepared by the Commonwealth Serum Laboratories and is highly effective.

In 1960, twenty-two-year-old Kenneth Earnest was bitten by a captive tiger snake three feet long. It belonged to his father and was used for display in a Buena Park, California, reptile exhibit. Kenneth had lost the use of his right eye in a childhood accident and apparently did not notice that the ill-tempered snake had moved close to its cage door when he went to feed it.

The Australian taipan *Oxyuranus scutellatus* grows to be 11 feet long. It is one of the largest venomous snakes in the world. It is generally acknowledged to be one of the most dangerous of all snakes. It is found in northern Australia and parts of New Guinea. There are no other species in the genus.

The African puff adder, *Bitis arietans,* is the most troublesome snake on that continent in terms of human injury. A large, common and important venomous snake.

The handsome Siamese or monocellate cobra, *Naja naja kauothia*, is a subspecies of the common or Indian cobra. This species is widespread throughout Asia and Africa and in some areas survives in large numbers.

The West African green mamba, *Dendroaspis viridis,* is a large, swift, active and sometimes aggressive snake that is much feared. It, too, however, is the subject of legend and unreasoning dread. Its bite can be survived and it regularly is. Intense advanced medical care is warranted, though, as soon as possible after the accident occurs.

The black mamba, *Dendroaspis polylepis,* is one of the largest, most aggressive, and most seriously venomous snakes in the world. Lengths to 14 feet have been suggested as not uncommon.

Within minutes of the bite he was gasping for breath and suffered from a violent headache. Tiger snake antivenin was obtained from the San Diego Zoo and Ken ended up in an iron lung. He became almost totally paralyzed. He received what has been described as superlative medical care from Dr. Findlay E. Russell. Neurologist Russell is a world authority on venomous injuries. Two weeks after the bite the young man could move his eye and follow his doctor's finger. He survived and is said to be the first person outside of Australia to live through a bite from one of these exceedingly dangerous animals. Whether or not the tiger snake is the "deadliest of all land serpents" as stated by *Life* Magazine at the time is open to some debate. It is, however, very high up on the list.

The extremely dangerous death adder (*Acanthophis antarcticus*) looks much more like a viper than an elapid, yet an elapid it clearly is. A smallish snake, seldom achieving three feet, it delivers an enormously effective bite with human fatalities said to have been at least 50 percent before antivenin was available. That mortality rate is extraordinarily high. Symptoms of a bite include vomiting, drowsiness, faintness, sweating, circulatory difficulty, a stumbling, staggering "drunken" gait, and difficulty in swallowing. The pupils of the eye are dilated, there is a clotting of the blood, and death usually ensues from respiratory failure. A death adder antivenin is prepared by the Commonwealth Serum Laboratories and recovery without its use is doubtful at best.

The various Australian brown snakes (genus *Demansia*, but also commonly *Pseudonaja* or "false cobra") are responsible for more snakebite deaths in Australia than any other snakes probably due in large part to the defensive propensities of the common brown snake *Demansia* or *Pseudonaja textilis*. All told, there are 17 whip and brown snakes in Australia and all are lumped into a single genus by at least some investigators. They are often abundant and wide-ranging in farming districts.

The mortality for human bite victims of the common brown snake is low—probably under 10 percent—but the snakes are common and encounters more often recorded than with most of the other dangerous species combined. The onset of symptoms is slow. Several hours may elapse between the bite and any symptoms except shock and fright. Eventually, though, there will be headache, dizziness, and weakness. Death can come from either cardiac or respiratory failure. Care must be taken since the typical symptoms of psychological shock—such as may be displayed by the victim of any snakebite—are similar to the early stages of actual brown snake intoxication. A specific antivenin is prepared by the Commonwealth Serum Laboratories and should be used in every case even before the onset of bite symptoms.

Australian tiger snakes *Notechis scutatus*, the most dangerous snakes in southern Australia. Systemic reactions to the bite are swift and extremely hazardous. It is a nocturnal animal and the danger arises from people inadvertently stepping on them.

The Australian brown snake *Demansia textilis*, an agile and sometimes aggressive species. It is probably responsible for more human fatalities in Australia than any other snake there. It will strike again and again if angry.

The exceedingly dangerous death adder *Acanthophis antarcticus* from Australia. The mortality rate from the bite of this small but powerfully toxic animal may be ten times that from the Indian cobra. It may be as high as 50 percent.

The variably colored five to six foot copperhead of Australia (*Denisonia superba*) (the Solomons copperhead is *D. par* and is about half that size) is a sluggish and retiring snake of Tasmania and the southern coastal region of mainland Australia. Bites are not common, but are serious clinical events. Deaths are uncommon. The bite syndrome consists of vomiting, rapid loss of muscle tone, and possibly coma. There is some blood involvement and there may be some circulatory distress. No antivenin specific to the copperhead is produced, but the tiger snake antivenin of the Commonwealth Serum Laboratories is commonly used and is apparently effective. It can be distressing to stumble into a copperhead since there are likely to be more nearby. They congregate in large colonies, at times, and are among the last snakes to go into hibernation and the first to emerge. Despite the name they have nothing whatsoever to do with the copperhead of North America which is a highly advanced pit viper of the genus *Agkistrodon* (also *Ancistrodon*).

The rough-scaled snake (*Tropidechis carinatus*) seldom exceeds 30 inches and has not been implicated in many human fatalities (perhaps no more than one). The snake is retiring and seldom aggressive. No antivenin is prepared to counteract its bite and tiger snake serum is said to be effective.

The genus *Pseudechis* is a source of some concern for people of the region where it is found. *Pseudechis australis*, the Australian mulga snake, also known as the king brown snake, is an active and sometimes aggressive five to six foot snake that will occasionally hang on once it bites and chew in a healthy (or *unhealthy!*) dose of venom. Human fatalities from this snake's bite, although rare, have been known. No antivenin is manufactured for the mulga snake, but several others produced by the Commonwealth Serum Laboratories are useful in the clinical picture: Taipan, tiger snake, and Papuan blacksnake antivenins all are used.

The red-bellied blacksnake (*Pseudechis porphyriacus*) grows to a length of eight feet and puts on a fine display when threatened. Raised like a cobra it feints and jabs, but its behavior is largely bluff. The bite, although potentially serious, is rarely if ever deadly. Symptoms are largely local and paralytic effects are not reported. There will be nausea and vomiting and local blood vessel involvement. No specific antivenin is produced, but Papuan blacksnake antivenin is useful.

The Papuan blacksnake (*Pseudechis papuanus*) is not very well known, but it is apparently more dangerous than its congener. A specific antivenin is produced by the Commonwealth Serum Laboratories. It has been suggested that there are more deaths each year resulting from the bite of this species than from the bite of any other snake from Papua, Australia, or New Guinea. There is no doubt that it is a dangerous snake.

The Australian copperhead *Denisonia superba* is in no way related to the North American snake with the same common name. This elapid is rather sluggish but will strike out if molested. Accidents are rare but serious. The specimen here is eating a green bell frog, *Hala aurea*.

Milking the red-bellied blacksnake, *Pseudechis porphyriacus*. This is the best known and most common of all venomous snakes in Australia. It puts on a great display if molested but bites are relatively few and the human mortality rate is probably no more than one percent. The venom in the glass, of course, comes from many specimens and not just this snake.

These, then, are the seriously venomous elapid snakes of the Australian region. Although there are some mildly venomous colubrids of the rear-fanged type, the elapids have the world down under to themselves. All told, 63 species of snakes in Australia are venomous, and of these 13 are seriously so. The figure of 63 accounts for almost 70 percent of the snake fauna of Australia. No other area on earth has anything approaching that proportion of venomous snakes. Although, once again, this fact hardly justifies hysteria it at least helps to explain the dread Australians have of their snakes. Snakebite deaths are few (from five to ten a year) and this is significant because the Australians are an outdoor people. As farmers, ranchers, and sportsmen they have ample opportunity to encounter their venomous snakes, but do so with far less frequency than the number of species might seem to indicate.

I have personally known a number of Australians with outdoor experience and their general lack of reliable information is surprising when it comes to their snakes, particularly since some of them are so exceedingly dangerous. I went to college with one

The mulga or Australian black snake (*Pseudechis australis*) can be aggressive, and when it bites it hangs on and chews to get more venom into the wound. Still, human deaths are rare and most accidents occur when snakes are captured or otherwise molested.

Australian chap who had spent most of his youth camping, fishing, and hunting. He has since become a well-known novelist and can only be described as extremely intelligent. He once rounded a corner and bumped into me while I was holding a rather large indigo snake and he very nearly had to be revived. He stumbled away in a near faint and later told me some of the most incredibly inaccurate stories about snakes I have ever heard. He insisted that there are no harmless snakes in Australia. A well-known Australian communications figure recently told me stories that contained almost as much nonsense. Such cases seem to be common. They are not cases of people deliberately lying or even exaggerating for effect. There just seems to be a wealth of misinformation in circulation despite some very good Australian books on the subject. Once again we would suggest that Australia has quite enough to worry about in relation to its venomous animals without enlarging the picture.

ASIAN ELAPIDS

In all of Asia there are only six genera of elapid snakes. Australia, New Guinea, and the Solomons alone have 23. However, the Asian genera cause some of the most intense snakebite problems in the world. Following the geographical analysis given by Henriques and Henriques (1971), the regions and the genera represented in this area of the world are as follows:

DISTRIBUTION OF ASIA ELAPID SNAKES

Region	Territory Included	Genera Represented
Western Asia	Turkey, Cyprus, Asiatic Egypt, Israel, Jordan, Lebanon, Syria, Saudi Arabia, Yeman, Handramat, Iraq, Iran, Transcaucasian U.S.S.R.	*Naja* (2)* *Walterinnesia* (1)
Central Asia	Afghanistan, Pakistan, India, Sri Lanka, Kashmir, Nepal	*Bungarus* (6) *Calliophis* (4) *Naja* (1) *Ophiophagus* (1)
Eastern Asia	Tibet, China, Korea, Japan, Ryukyu Islands, Formosa	*Bungarus* (2) *Calliophis* (4) *Naja* (1) *Ophiophagus* (1)
Southeastern Asia	Burma, Thailand, Laos, Cambodia, Vietnam, Malaysia, etc.	*Bungarus* (6) *Calliophis* (3) *Maticora* (2) *Naja* (1) *Ophiophagus* (1)
Indonesia	Sarawak, Brunei, Northern Borneo, Indonesian and Philippine Islands	*Bungarus* (4) *Calliophis* (2) *Maticora* (2) *Naja* (1) *Ophiophagus* (1)

*Number of species.

The very words "venomous snake" conjure up in most minds the hooded cobra, forebody raised off the ground, weaving back and forth waiting to strike. The cobra is sacred in many parts of its range. In India a day is set aside in July and another in October for the veneration of this serpent. Throughout Hinduism and other Asian religions as well the cobra appears again and again as a god, goddess, messenger, or agent of the supernatural. Temples have been built for the Naga or "cobra kings." Again and again the theme is repeated: The cobra is sacred and imbued with magical powers.

During my visits to Sri Lanka I spent time with herpetologist Ranil Senanayake. He owned at that time the only known albino cobra. It was not a very large snake, under three feet, but it was white with shocking pink eyes. A regular procession of Buddhist monks appeared at Ranil's door for as long as he owned the snake, coming from towns and villages all over the country. They came as pilgrims to claim the serpent as the rightful property of their temple. They presented long and involved stories of how the coming of the sacred white cobra had been foretold, always in relation to their own particular temple, and how, as a messenger from Lord Buddha, the snake must go with them to dwell in that temple. The visitors did not stop coming until well after the news of the snake's death, after several years of captivity, had reached the farthest corners of the land.

The common or Indian cobra, *Naja naja.* This is a common, prolific, adaptable snake found over much of Asia in the form of one subspecies or other. They can be aggressive but their venom is nowhere near as powerful as fiction writers would have us believe. Human mortality probably doesn't exceed five percent. These snakes are common around human habitations where they hunt for rats and accidents are frequent.

There are six species of cobra in the genus *Naja* (pronounced neye-ah) five of which are African. Only one full species is found in Asia*—*Naja naja*, the common, Asiatic or Indian cobra. It has many subspecies and a vast range. The principal subspecies are currently recognized as:

Naja naja naja	Asiatic cobra
	Peninsular and Northern India,
	Sri Lanka, southeastern Pakistan
Naja naja oxiana	Oxus cobra
	Northern Pakistan, Afghanistan,
	Iran, Russian Asia
Naja naja kaouthia	Monocellate cobra
	Eastern India, Bangledesh, Assam,
	Burma, Thailand, undetermined
	range in China and Malaya
Naja naja atra	Chinese cobra
	Thailand, China, Vietnam, Taiwan
*Naja naja sputatrix**	Malay cobra
	Malay Peninsula, larger islands
	of Indonesia
Naja naja miolepis	Borneo cobra
	Borneo, Philippines
Naja naja samarensis	Visayan cobra
	Philippines
Naja naja philippinensis	Philippine cobra
	Philippines
Naja naja sagittifera	Andaman Island cobra
	Andaman Islands
Naja naja sumatrana	Sumatran cobra
	Sumatra

*This species is apparently capable of spitting its venom, a phenomenon we will discuss when we come to the Elapidae of Africa.

The cobras are so widespread and of such ongoing concern to man that antivenin for the Asiatic forms is now produced in ten countries: Germany, India, U.S.S.R., Thailand, Australia, Taiwan, Iran, Philippines, Indonesia, and France.

Cobras in general are shy snakes and prefer to retreat rather than face a showdown when on the defensive. Still, they are common around human habitations and cultivated areas (where they hunt for rats and frogs) and encounters occur often enough for a vast number of episodes to occur.

The four to eight foot cobra is nocturnal in many parts of its range, but in some other parts, notably India and Pakistan, it tends to be diurnal. The reluctance of this snake to strike is nowhere better seen than when snake catchers are afield. With ease they grab previously unhandled cobras by the tail and carry

*This long established "fact" may be open to some question.

them around. Handlers and catchers are on occasion bitten, but their skill and speed is usually greater than the snake's.

Cobras seem reluctant to use very much of their venom when striking out in defense. If they are seriously injured or wildly infuriated, it may be another matter. Venom yield is not necessarily related to snake size, but it has been demonstrated that a good sized cobra can yield 600 milligrams of venom when milked in the laboratory. If this amount were delivered in a defensive strike, the recipient would get about 30 times the amount necessary to kill him—calculated to be about 20 milligrams for a full-grown man.

How deadly is the cobra? To begin with the legends are pure nonsense. Ahuja and Singh (1956) did a study of reported snakebite cases in India. The encounters occurred between 1940 and 1953 and as reported, of course, represent only a small fraction of those that took place. They reported on 280 cases of elapid envenomation—which resulted in 77 deaths. Comparing the Asiatic cobra with the much more virulent Indian krait (*Bungarus caeruleus*), they found the following:

	Number of Bite Cases	Number of Deaths	Mortality
Naja naja	131	11	8.4%
Bungarus caeruleus	35	27	77.1%

Chaudhuri and his co-workers (1971) suggest that in India overall mortality from snakebite may run between 30 and 40 percent. Obviously, the cobra with its suggested mortality of 8.4 percent could be responsible for only a part of these. So much at least for the deadly cobra "whose bite means certain death." (The figure of 30-40 percent seems much too high and is almost certainly incorrect.)

In one series of cobra bite cases in Malaya, Reid (1963) reported on 33 clinically-treated patients. Of the 33, 20 had received no venom at all although bitten by cobras. Of the 13 that were envenomated, five were listed as either trivial or moderate and three were classified as severe. One death occurred in the series. There is a strange character to cobra bites in Malaya. Although laboratory animals injected with cobra venom from species taken in Malaya display the usual elapid paralytic syndrome, this element of intoxication is usually not seen in the clinic. There is often considerable tissue damage, usually deep within the tissues in the locale of the injury. Of the 33 cases in the Reid series, only four showed any sign of paresis or partial paralysis. And where paresis is seen in Malayan cobra bites it is rapid in onset, but short-lived. The low level of neurotoxic involvement in Malayan cobra envenomations has not been properly quantified or explained.

Sometimes mistaken as nothing more than an overgrown cobra is the king cobra, *Ophiophagus hannah*. Also called the hamadryad, it is clearly one of the most dangerous snakes in the world. Diurnal in habit and usually found only in remote areas away from human habitations, the king cobra is only rarely involved in serious encounters with human beings. It is the giant among venomous snakes, easily the largest in the world. Sixteen to 18 foot specimens are recorded. Although 19 feet has often been quoted as a maximum length, I have never seen a verified account of a specimen that length (18′4″ is given by Minton).

The king cobra spreads its hood, rather like the *Naja* cobras except that in relation to body size the king cobra's hood is small. Many people are confused by the cobra's hood. It is not a permanent fixture, that is, cobras "don't go around looking that way." In movement from place to place or in repose the cobra displays no hood at all. Only as a threat display does the cobra rear and spread the ribs *behind the head* creating its famous hood.

The inevitable picture—mongoose versus Indian cobra. These fights do occur and the mongoose does prey on the cobra as well as other snakes, lizards, birds, and small mammals. The mongoose is not immune to cobra venom and a misjudgment could mean the predator becomes the victim.

The king cobra is a snake-eater and is nowhere plentiful. Its range is vast—Peninsula India to the foothills of the Himalayas, eastward across southeast China into regions to the south, the Philippines, the larger Indonesian Islands, Malaya, and Thailand. Of course, in great areas of this range there will be no king cobras at all, much of the suitable land centuries ago having been taken over by human inhabitants.

The olive, brown, or greenish yellow king cobra has a peculiar habit not seen in most other snakes. The female actually constructs a nest by using her muscular body to pull together a pile of leaves and other forest debris. Into the middle of the heap she

The business end of the king cobra (*Ophiophagus hannah*). This, the largest venomous snake in the world, causes relatively few accidents because of its tendency to avoid human habitation. It is a snake-eating snake.

lays her eggs and then *guards them.* Coiled in, around, or near the nest, she protects it from intruders. Stories differ as to how determined she is in her efforts to protect her eggs, but it may be that the bites that do occur result from people inadvertently approaching nests.

Although an antivenin for the king cobra is produced by the Queen Saovabha Memorial Institute in Bangkok, survival from the bite is problematical. The bite is not inevitably fatal, but the mortality rate is apparently extremely high. The venom, drop by drop, may be less toxic than those of a number of other elapid venoms, including that of the common cobra, but the quantity given by this enormous animal can be extremely dangerous.

Reid (1968) discusses king cobra bite victims. A Burmese snake-catcher was bitten by a ten-foot specimen and was not envenomated, while a Chinese handler was bitten by a 16-foot specimen and suffered only slight local necrosis, no systemic effects, and was healed and well within two weeks. Reid also discusses briefly a case involving a fifty-eight-year-old Malay farmer who sat on a 12-foot king cobra and died six hours later after being bitten on the buttock.

In comparing the lethal dosages of elapid venoms administered to laboratory animals intravenously, Kocholaty and his co-workers (1971) found that the following species (among many others tested) were more lethal than the king cobra (on a drop-by-drop of venom basis) by the factors shown:

Bungarus multicinctus
Many-banded krait   22.53 times as deadly

Bungarus caeruleus
Indian krait   14.11 times as deadly

Micrurus fulvius
North American coral snake   4.23 times as deadly

Naja naja naja
Common or Asiatic cobra   6.55 times as deadly

Naja naja naja
Monocellate cobra   4.49 times as deadly

Still, not one of these other species could deliver anywhere near as much venom as the king cobra. One rough estimate has this snake capable of delivering 120 times the amount needed to kill a man.

Ganthavorn (1971) states that there is not a single case on record of a king cobra bite occurring "in nature" in Thailand, yet the Queen Saovabha Memorial Institute does produce the antivenin for this species. In her paper she reports the case of an employee at the Institute who was bitten while handling a king cobra. I have on a number of occasions watched the handlers at the Institute in Bangkok and it is a mystery to me how any of them survive at all. They walk around in enclosed snake "pits" with 20 and more king cobras around them. In sorting them out for milking, feeding, examination, or just public display they pick up two snakes at a time, balancing them at approximately the center of their bodies, and put them into bins or release them to swim in the small, shallow moats around the pits. Snakes rearing as if to strike are simply pushed aside with little more than a glance. It is quite the most spectacular snake-handling display I have ever seen.

As reported by Ganthavorn, a fifty-eight-year-old handler with 25 years' experience misjudged his snake at 11:30 A.M. one day in January, 1970. He was approaching the snake to force-feed it when the snake turned and bit him between the thumb and index finger on his right hand. Ganthavorn specifies only that the snake was "huge."

The victim received initial treatment in the emergency room at the Institute. A tourniquet was applied, and he was given a large dose of specific antivenin (prepared in the next building) and cortisone, to prevent serum reaction.

Within a few minutes of the bite the patient experienced local pain and swelling. Thirty minutes later he felt sleepy and his eyelids drooped. Pain and swelling at the bite progressed. Seventy minutes after the bite he experienced difficulty in breathing and swallowing. He finally lost consciousness and was taken to Chulalonghorn Hospital. An endotracheal tube was inserted and he was placed in a respirator as his breathing became progressively more labored. Periodically he would stop breathing altogether and the respirator was essential to maintain life. His pupils did not react to light and there were no reflexes in the arms or legs. The most critical symptoms developed within a very short period of time, about 15 minutes from onset. He was given more antivenin, cardiac stimulants, and hydrocortisone after he reached the hospital.

Three hours after the bite the victim was still unconscious and having difficulty breathing. His blood pressure had dropped and was irregular. It was not until eight hours after the bite that his pupils began to react to light and some movement was noted in one leg. He remained unconscious and still stopped breathing intermittently. More antivenin was given. Twelve hours after the bite symptoms remained the same.

The crisis passed after 19 hours and the patient regained consciousness. It was not until that point that the respirator was removed and the administration of antivenin stopped.

Four days after the bite the swelling in his arm—which had extended from his hand to his upper arm—began going down and the pain began to subside. The fang wounds themselves became seriously infected (*Proteus vulgaris*) and it was not until a month after the bite that he could be released from the hospital. All told, 1,150 milliliters of serum had been administered in saving this man's life.

The detailed recounting of this episode here is to stress the kind of care that was required to save the victim's life, and that, apparently, only barely. He could not have lived, it is clear, without massive doses of antivenin and without the respirator. If one must be bitten by a king cobra at all it is certainly advisable for it to happen on the grounds of the institute where these snakes are studied and their venom taken for production of antivenin. A barefoot farmer in a remote Asian village would probably not fare as well.

A third genus of elapid found in Asia is limited to a single species, a cobra-related animal. *Walterinnesia aegyptia*, the desert black snake, is found from Egypt to Iran. It is known to achieve a length of four feet and although little research has been done on it to date it is presumed to be dangerous. It is assumed that the venom is about as toxic as that of the Asiatic cobra, although less is apparently administered in a bite. No antivenin is prepared.

There are three more genera of elapids in Asia and we will touch on the two of least concern before going on to discuss the kraits.

Calliophis is represented by 13 species, all from Asia and all variously known as Oriental coral snakes. Encounters involving human beings are rare and it may be, as is commonly thought, that these snakes themselves are rare. It may also be that they are more common than is supposed and it is only their retiring nature that keeps them out of trouble with man. Some few of these snakes may exceed three feet although they would be exceptional specimens. It is generally assumed that the larger specimens are dangerous on those rare occasions when a bite might occur. It is assumed that a bite is most unlikely unless the snake is being handled. No antivenin is prepared.

The *Maticora* are the long-glanded coral snakes and have a bizarre adaptation. Their venom glands are grossly enlarged, or at least elongated, and run a third the length of the animal's body. In other snakes the venom glands are restricted to the area of the head and neck. This special feature in the *Maticora* has required the movement of the heart down to about the center of the animal's body. Bites are rare, but not unknown. A bite by a specimen of *M. intestinalis*, the belted coral snake, is known to have caused serious illness, while a case involving *M. bivirgata*, the white-striped coral snake, is known to have caused a human death. The genus is limited to southeastern Asia and Indonesia. There is no antivenin available.

Chaudhuri and his co-workers (1971) attempted to calculate the fatal dosages for man of various elapid venoms. Figuring their doses in milligrams they postulated the following:

Lethal Dosages for Man—Asia Elapids

Naja naja common cobra	15.0
Ophiophagus hannah king cobra	12.0
Bungarus fasciatus banded krait	10.0
Bungarus candidus common krait	1.0

This should make it clear why the kraits, genus *Bungarus*, are of special concern to man in Asia. The very deadly common krait possesses a venom 15 times as powerful as that of the cobra. Even the considerably less lethal banded krait takes a third less venom than the cobra to kill a man. This is substantiated by the findings of Ahuja and Singh in 1956, it will be recalled, when they stated the mortality rate of the Indian krait (*B. caeruleus*) to be 9.17 times as high as that of the cobra.

The genus *Bungarus*, encompassing all the kraits, is found only in Asia. There are probably 12 species in all. They are usually given as follows:

THE ASIAN KRAITS—GENUS: *BUNGARUS*

Bungarus bungaroides	Upper Burma, India, Sikkim
Bungarus caeruleus Indian krait	India, Pakistan, Sri Lanka
Bungarus candidus Malayan krait	Thailand, Vietnam, Malaya, Sumatra, Java, Celebes
Bungarus ceylonicus Ceylon krait	Sri Lanka
Bungarus fasciatus banded krait	India, Bangledesh, China, Burma, Thailand, Cambodia, Laos, Vietnam, Malaya, Sumatra, Java, Borneo
Bungarus flaviceps red-headed krait	Thailand, Vietnam, Malaya, Sumatra, Java, Borneo
Bungarus javanicus Javan krait	Java
Bungarus lividus	India, Bangladesh
Bungarus magnimaculatus	Burma
Bungarus multicinctus many-banded krait	China, Burma, Taiwan
Bungarus niger black krait	Northeast India, Sikkim
Bungarus walli	Eastern India

The Indian krait (*Bungarus caeruleus*) in a typical pose, head concealed. The venom of this snake is exceedingly toxic to man and estimates of mortality rate run from 77 percent to as high as 89 percent. This species is active at night and is relatively lethargic during the day.

The banded krait (*Bungarus fasciatus*) is a quiet and inoffensive snake and it takes considerable prompting to get one to bite. It is so peaceful, in fact, that many people consider it harmless although it is clearly a venomous snake.

The vividly marked, smooth-scaled kraits are harmless-looking snakes offering none of the dramatic display of the cobras, the rattlesnakes, or the threatening elapids of Australia. They are nocturnal and may display markedly different behavior during the daylight hours and after dark. During the day many species, at least, tend to be sluggish and inoffensive. At night, however, on the prowl for food they can be more aggressive and therefore dangerous to a barefoot person moving along a trail. Even the most inoffensive snake can take exception to being stepped on. In general, though, bites by the various kraits (that is pronounced *krite* not the usually heard *crate*) are rare. When they occur, however, the mortality rate is exceedingly high, for some species it is among the highest in the world.

Antivenins specific to the genus *Bungarus* are produced in a number of places, so good serum therapy is theoretically available anywhere the snakes are found. In fact, though, lack of facilities, lack of communication, and confusion as well as superstition keep victims in remote areas from receiving the antitoxin which in the case of some kraits is virtually essential to survivial. *Bungarus* antivenins are now produced by:

Institut Pasteur, Paris	*B. flaviceps*
Central Research Institute, Punjab	*B. caeruleus*
Haffkine Institute, Bombay	*B. caeruleus*
Perusahaan Negara Bio Farma, Djalan	*B. fasciatus*
Institut d'Etat des Serums et Vaccins, Teheran	*B. fasciatus*
Taiwan Serum and Vaccine Laboratory, Taipei	*B. multicinctus*
Queen Saovabha Memorial Institute, Bangkok	*B. fasciatus*

In the case of some species, notably *B. caeruleus* and *B. candidus*, treatment without the use of appropriate antivenins can be tantamount to no treatment at all.

Minton and Minton (1969) give estimated lethal doses of venom for an adult human. By extracting a few of these it is possible to see at least one of the more dangerous kraits in context. The doses given here are quoted directly from an appendix to their *Venomous Reptiles*:

Boomslang *Dispholidus typus*	"Very small, probably 5 mg. or less"
Death Adder *Acanthophis antarcticus*	"About 10 mg. . . ."
Australian Brown Snake **Pseudonaja textilis*	"Very small, probably 2-3 mg."
Tiger Snake *Notechis scutatus*	"About 3 mg."
Indian Krait *Bungarus caeruleus*	"Very small, probably 2-3 mg. About 50% of bites fatal even with antivenin treatment."
Eastern Green Mamba *Dendroaspis angusticeps*	"About 15 mg."
Asian Cobra *Naja naja*	"About 15-20 mg."
Puff Adder *Bitis arietans*	"Lethal dose 90-100 mg."
European Viper *Vipera berus*	"Lethal dose 20-25 mg."
Cottonmouth *Agkistrodon piscivorus*	"Lethal dose 100-150 mg."
Western Diamondback Rattlesnake *Crotalus atrox*	"Lethal dose believed to be about 100 mg."

*Given elsewhere in this book as Demansia.

One of the kraits, the Javan (*Bungarus javanicus*), is believed to be rare. It is found only on the north coast of Java and is said to be small, rarely over 38 inches. Although little is known of its toxicity, it was said to be a representative of this species that crawled into the hut of a sleeping family and bit a man and his adult son. Both men are said to have died. This would make the species dangerous indeed.

P. J. Deoras, who has spent a lifetime in India working with venomous animals and their toxins, in his paper (1971) said: "The symptoms of krait poisoning are similar but milder than with cobra envenomisation [sic] and there is little local swelling or pain. There is also no nausea or frothing at the mouth. The victims sleep and gradually die after the numbness in the extremities extends to the neck region. There is also albumin found in the urine."

There is a great range of toxicity among the kraits. One study, at least, (Kocholaty *et al.*, 1971) would seem to indicate that one krait venom is 12.41 times as powerful as that of one congener and 19.40 as strong as that of another.

The venoms of the various kraits, when dried, show themselves to be about 90 percent protein with the balance inorganic material. There are powerful neurotoxic factors, enzymes, and hemolysin present. The relative quantities of these materials in the different species is apparently not yet a matter of record. In clinical reports the local reactions are always slight to nil. Death times, not an academic matter when discussing intoxication by the krait, are moderately rapid. In one series (reported by Reid, 1968) 18 fatal cases were analyzed. The average elapsed time from bite to death was 18 hours. The range was from three to 63. The species was *B. caeruleus*. It is interesting to compare the average of 18 hours to the average in 27 cases of *Naja naja* bites—8.4 hours.

While the cobra may be considered the sacred or mystical snake of Asia and, with its hood spread, the symbol of venomous creatures, the lesser-known kraits are the recipients of the horror stories. It is a krait, called the "seven stepper" for reasons explained elsewhere, that the traveler is told is inevitably and instantly fatal. While it is true that some *Bungarus* species do have an appalling mortality rate to their credit, bites are still a rarity. The kraits remain quiet, seldom aggressive, and even shy snakes. When they do strike, however, the recipient of their wrath is likely to be seriously and even critically ill within a short time. Without serum therapy there is a strong likelihood the victim will succumb, if venom has been introduced into the wound.

In the next chapter we will discuss the elapids' presence and diversity on the great African continent. We are not through with Asia, however, for we must return twice more—with our discussions of the Viperidae and Crotalidae. It will be only after introducing these two additional groups of snakes that we will be able to understand even a little of the snakebite problem in that vast and heavily populated area of earth.

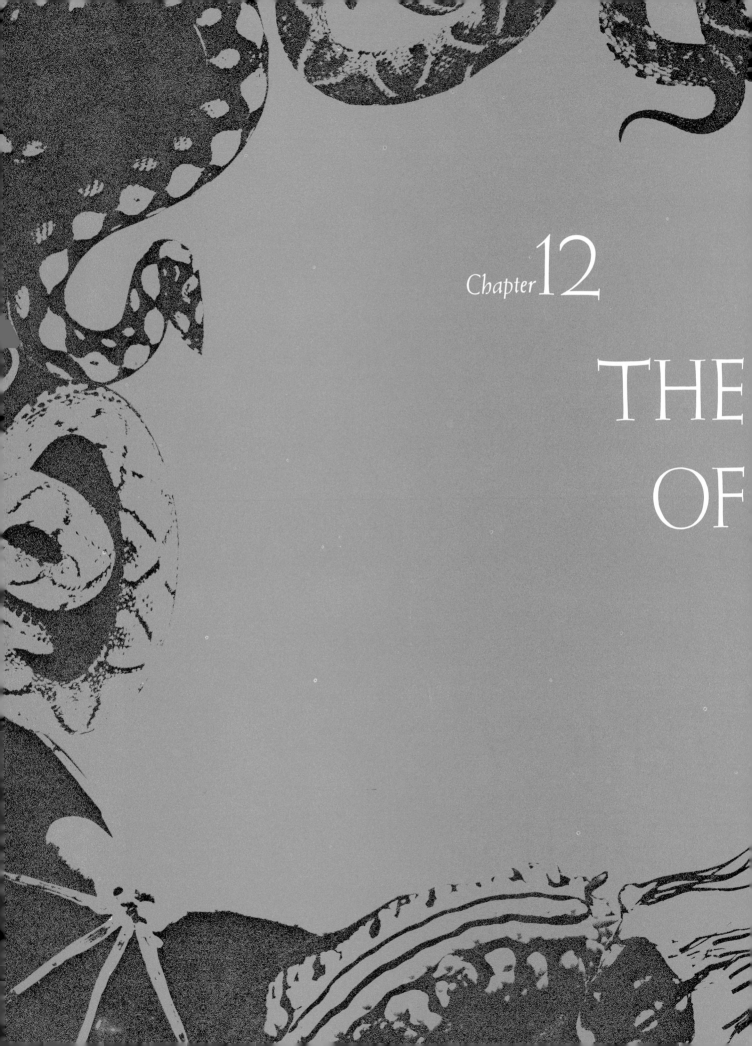

Chapter 12

THE
OF

ELAPIDS

AFRICA

The Egyptian cobra (*Naja haje*) is a widespread, dangerous species and accounts for a number of human fatalities each year. It can be very aggressive and will move around human habitations in search of rats. It will enter occupied buildings.

he elapids are well represented on the vast African continent. Again using Henriques and Henriques' (1971) geographical divisions, we find the following genera in the areas indicated:

DISTRIBUTION OF AFRICAN ELAPID SNAKES

Region	Territory Included	Genera Represented
Northern Africa	Spanish Sahara, Morocco, Algeria, Tunisia, Libya, African Egypt	*Naja* (1) *Walterinnesia* (1)
Central Africa	Mauritania, Sénégal, Gambia, Portuguese Guinea, Guinea, Sierra Leone, Liberia, Ivory Coast, Mali, Ghana, Togo, Dahomey, Nigeria, Chad, Cameroon, Ubangi, Spanish Guinea, Sao Tomé Island, Gabon, Cabinda, Congo, northern Angola, Ethiopia, French Somaliland, Somalia, Uganda, Kenya, Ruanda-Urundi, Tanganyika, Zanzibar, northern Mozambique, northern Rhodesia	*Boulengerina* (2) *Dendroaspis* (4) *Elapsoidae* (1) *Hemachatus* (1) *Naja* (4) *Paranaja* (1) *Pseudohaje* (2)
Southern Africa	Southern Rhodesia, South West Africa, Bechuanaland, South Africa, southern Mozambique	*Aspidelaps* (2) *Dendroaspis* (2) *Elaps* (3) *Hemachatus* (1) *Naja* (5) *Pseudohaje* (1)

There are no Crotalidae in Africa at all and discounting the two venomous rear-fanged colubrids we have already discussed, all the venomous land snakes of Africa are divided between the Elapidae and the Viperidae. The number of genera and species from each family are:

Area	Number of Genera		Number of Species	
	Elapidae	Viperidae	Elapidae	Viperidae
Northern Africa	2	4	2	7
Central Africa	7	7	15	28
Southern Africa	6	3	14	11

It is obvious that the elapids must be examined with care if one is to understand the snakebite problem anywhere in Africa. They have developed well on that vast land mass, they have diversified, and in each situation they have found ways to survive in the complex interaction of man and animals. From dense forests to deserts, from highlands with bizarre and exotic fauna and flora, to steaming lowlands, the elapids of Africa have made their way. They are a fascinating group of animals with powers as stunning as those of almost any other venomous snakes on earth.

MAMBAS

No snakes in Africa inspire more universal dread than those belonging to the genus *Dendroaspis*—the four species of mamba. It is not without reason that they are the most feared snakes on the continent. One professional snake-catcher from South Africa who advertises in the international journal on toxicology (*Toxicon*) offers pure venom for research and study. His prices may be indicative of the pervasive attitude toward the mambas:

Egyptian cobra (*Naja haje*)	$16.00/gram
Ringhals (*Hemachatus haemachatus*)	12.00/gram
Puff adder (*Bitis arietans*)	10.00/gram
Black mamba (*Dendroaspis polylepis*)	70.00/gram

As Christensen and Anderson (1967) comment: "Although the mambas, rightly or wrongly, are feared more than any other African snake, only little is known about their venom, presumably because nobody has been anxious to collect it. . . ."

The mambas are among the most dangerous creatures on earth. At least one, the black mamba (*D. polylepis*), grows to be 14 feet long, making it the second largest venomous snake in the world. This snake is extraordinarily fast and its venom is particularly toxic.

One series of snakebites in Natal between 1957 and 1963 was reported by Chapman (1968). There were 40 cases in all. Ten people died, a 25 percent mortality rate. Seven of the ten were the seven mamba bite cases in the series—a mortality rate of 100 percent. This may be compared with 14 cases of bites by the ringhals cobra in the same series—with no fatalities. It should be noted that the series was a recent one and sophisticated medical techniques would have been available. Still, black mambas managed to kill seven out of seven people.

Mortality from mamba envenomation does not run a consistent 100 percent, of course, but it is certainly true that anyone bitten by the black mamba is in serious trouble from the outset. A polyvalent antivenin for *Dendroaspis* is produced by the South African Institute for Medical Research in Johannesburg. Without it a bite victim's chances for recovery are slim.

The mambas inhabit most of Central and Southern Africa. They are widespread and in some places common. The four species are distinct in a number of ways:

> *D. angusticeps:* The Eastern green mamba has a relatively narrow range in the forests and brush lands of East Africa. It ranges southward from Kenya to the Cape Province in South Africa and also occurs on Zanzibar. It is arboreal and is found on the ground less often than the other mambas. It grows to be eight feet in length and although it is a very large snake its venom is only about half as toxic as that of the black mamba.

> *D. jamesoni:* Jameson's mamba is another green tree snake. It is essentially an animal of the tropical rain forests and is found from western Kenya and Tanzania to Guinea and Angola. Adults have been known to reach eight feet though admittedly that would be an outside figure. Although arboreal, this snake would be more likely to be encountered on the ground than the Eastern green mamba.

> *D. viridis:* The Western green mamba, as the name implies, is the opposite number to *D. angusticeps.* This seven-foot arboreal snake ranges through the rain forest areas of Africa's western bulge: Senegal to the Niger, and the island of Sao Tomé. Not too much is known about the snake but, reflecting their interest in the area, the French produce a monovalent antivenin for *D. viridis* at the Institut Pasteur in Paris.

> *D. polylepis:* The black mamba is the mamba that gives the others their evil reputation. It is not only the second largest venomous snake in the world, but also one of the fastest. It has been reliably clocked at seven miles an hour. (Claims of snakes dashing around at 25 and 35 miles an hour are sheer nonsense, of course.)

The eastern green mamba (*Dendro-aspis angusticeps*) is an active species seldom found on the ground. It stays in trees and bushes and can move through them at considerable speed. The black mamba is much more aggressive, though, and has a venom that may be twice as toxic as that of this species.

The black mamba is less arboreal than the other three mambas although it is commonly found in trees and thick bushes. It can move through foliage at least as fast as it can move on solid ground. Its size and aggressiveness, and the extreme toxicity of its venom, are compounded by the fact that the black mamba has one of the longest strikes of all venomous snakes. It is reputed to be able to strike between 25 and 40 percent of its length. The venom injected inhibits breathing and apparently attacks that branch of the vagus nerve involved in the regulation of the heart. Typically, in black mamba intoxication, there are periods of wild, erratic heartbeat.

It has been estimated that two drops of black mamba venom would be the lethal dose for an average sized man. A large black mamba has yielded 15 drops when "milked." Whether or not a mamba in an average biting situation (if such a thing exists) would discharge that impressive quantity is not known.

Dendroaspis envenomation is characterized by the speed with which the symptoms appear. All snake venoms affect the blood's ability to coagulate to some extent. The venoms of the mambas

Jameson's mamba (*Dendroaspis jamesoni*) is less known than the other mambas, but it is known that it frequents trees and ground haunts as well. It stays in tropical rain forests and seems to avoid dry areas.

The eastern coral snake (*Micrurus fulvius fulvius*), an unusual snake for North America with a venom that is essentially neurotoxic. Although seriously venomous, it comes nowhere near living up to its reputation of being inevitably lethal.

From Nigeria the black-necked spitting cobra, *Naja nigricollis*, one of the most accomplished of all spitters or squirters of venom. Accuracy and range claims for all venomous snakes that squirt their venom are grossly exaggerated. This snake might be reasonably accurate at six or seven feet but probably not more than that.

The common cobra, *Naja naja*, in the exceedingly rare, pure albino form. This specimen was photographed in Colombo, Ceylon, and was the subject of controversy as long as it lived in captivity—a little over four years. It was proclaimed a sacred messenger from Lord Buddha and demanded by many monasteries.

Naja nivea, the Cape or yellow cobra seen in the golden phase. Ounce for ounce this cobra produces one of the most seriously toxic venoms of all the cobra group.

are not exceptions. The action of black mamba venom, at least, is variable according to dosage—it has been called "double action." In low concentrations the venom is an anticoagulant. In high concentrations it is a coagulant. But, although the blood involvement of mamba venom could have profound long-range effects, the matter of such effects is usually academic. Long before circulatory shock can develop the victim of the bite is suffering profoundly from respiratory distress.

There are few local effects in a mamba bite. Seldom is there any suggestion of pain, and swelling is usually absent. There will be profuse salivation and some dizziness. Restlessness and psychological disturbance may be followed by unconsciousness and a drop in blood pressure. The patient always displays difficulty in breathing. Some allowance must be made for the psychological reaction of a patient if he knows he has been bitten by a mamba. All kinds of symptoms could arise from the feeling of dread that would inevitably ensue.

Because the black mamba is so large and so very venomous there is the natural and perhaps inevitable tendency to create a

The mouth of a black mamba (*Dendroaspis polylepis*), one of the world's most dangerous snakes. This is a species many professional handlers perfer to avoid. They are fast, aggressive, very large, and exceedingly venomous. Lengths to 14 feet have been reported.

mental picture of the monster stalking human prey. That is, of course, wholly inappropriate. There is no doubt that the black mamba is a nervous snake. When encountered it may draw back, spread a slight hood, and open its mouth wide as a threat display. However, if not threatened it will often back down. One friend told me of being in the field with a companion in South Africa when they encountered a black mamba they estimated to be 12 feet long. It displayed after its fashion and the two boys froze. They had had experience with mambas before. After satisfying itself that it had been suitably terrifying, the snake slithered down from the bush where it had been when encountered and crawled across the feet of the boy nearest to it and between the legs of the other. It did not offer to strike either. They stood absolutely paralyzed and when they finally turned their heads to be certain the coast was clear they saw that the snake had ascended into another bush about ten feet away and was watching them very intently. The movement they made with their heads prompted the snake to do its threat display a second time. Only after several more minutes did the nervous mamba move away. There is no way of describing the condition of the nerves of the two teen-agers.

LESSER AFRICAN ELAPIDS

We will look next at the lesser elapids of Africa. Very little is known about either the habits or the venoms of these snakes.

Paranaja multifasciata, the burrowing cobra, is a little known two foot long snake with notably long fangs. It is an elapid and must be considered dangerous, or potentially so if handled. The species, the only one in the genus, is known from western Central Africa. There is no specific antivenin available for this snake.

Elapsoidea sundevallii, the African garter snake, is again the only species in its genus. It is found in 11 races over most of tropical and southern Africa except in the region of the Cape. It can grow to four feet in length and must be considered dangerous although it is characteristically sluggish and inoffensive. It is said to bite only in self-defense, and even then reluctantly. Little is known of its habits or venom and there is no antivenin specific to this animal.

Aspidelaps is a genus of two species—the African shieldnose snakes. They are small, semi-burrowing animals usually under 30 inches in length. They are restricted to the southern third of Africa and are not considered dangerous to man although they are elapids and therefore distinctly venomous. The two species are *A. lubricus*, the African coral snake, and *A. scutatus* the shield-snake. No antivenin is available and nothing, or nearly nothing, is known of the venom of this genus.

Elaps, another African elapid genus, contains two species: *E. lacteus*, the southern dwarf garter snake, and *E. dorsalis*, the striped dwarf garter snake. These two South African species, once again, are certainly venomous, but not considered dangerous except under bizarre circumstances. Nothing is known of their venom and no antivenins are available, or apparently needed.

In the tropical rain forest regions of Central and western Africa there are two tree cobras of the genus *Pseudohaje*. They grow to be as much as eight feet in length and are certainly both dangerous. Virtually nothing is known about their behavior or the chemistry of their venoms, but it is assumed that since both are elapids and both grow to a very large size they are capable of administering a seriously intoxicating bite. There is no antivenin available for the bite of *P. goldii*, Gold's tree cobra, or *P. nigra*, the black tree cobra.

Two other "cobras" with habits quite different from those of the tree cobras inhabit Central Africa. The genus *Boulengerina* contains the water cobras. They, too, grow to be eight feet long and are known to be seriously venomous. They are not generally aggressive and bites, while known, are relatively rare. The venom, as might be expected, is a powerful and rather fast-acting nerve toxin and symptoms are systemic rather than local. No antivenin for these snakes is produced. The two species are *B. annulata* and *B. christyi*, the banded water cobra and Christy's water cobra respectively.

In passing the lesser genera it is worth noting a comment by Christensen (1968): "There is nothing known about the composition, actions, and immunological properties of the venoms of Central and South African snakes belonging to the genera *Elaps, Paranaja, Pseudohaje*, and *Vipera*. . . ." This comment sums up a condition that is more often true of a venomous animal than not true. Information is meager, research time-consuming and necessarily both sophisticated and expensive. It will be a long time before anything like an analysis is completed of the lesser genera of venomous snakes. First attention must be given to understanding those species which figure on the public health statistical tables. It is safe to say, I think, that there are species of venomous snakes, even some belonging to as important a group as the Elapidae, that will become extinct before an analysis of their venom is undertaken. The kind of money and expertise required for learning more about these animals will always be tied up in research the direct benefits of which can be seen in advance—such as the reduction of human mortality. One feature of venom research that, although not unique, is always strongly apparent is its pragmatism.

The name *spitting cobra* is confusing. There is a spitting cobra in Asia (genus *Naja*) and another in Africa of the same genus (*N. nigricollis*), but the most highly developed of all spitting cobras goes by a different name, the ringhals of southern Africa. It is *Hemachatus haemachatus*, the only species in the genus. This snake grows to be five feet long and can be aggressive. Its venom is usually aimed at the eyes and although claims of accuracy are characteristically exaggerated strikes are commonly made and intense pain ensues. There are severe spasms of the eyelids and, unfortunately, the pain and after effects do *not* fade away in time. Unless a neutral flush is applied to the eyes very soon after contact, permanent damage to eye tissue and even blindness can result. Zoo workers are required by safety rules to wear a welder's face shield, or a version of it, when exposing themselves to this strange and intriguing animal.

It is usual for casual readers in natural history (and even some people who should know better) to concentrate on a few outstanding characteristics of each animal. Thus it is that the ringhals is thought of as an animal that spits its venom. It is not as often thought of as an animal that injects its venom through well-suited fangs. If challenged or threatened this snake will endeavor to incapacitate a real or imagined foe by blinding it, but that is not how it kills. It does bite people often enough for there to be an unusual number of antivenins available to repair the damage. Two serums are produced by the South African Institute for Medical Research and one by CAPS Put. Ltd. in Rhodesia. Protection is also afforded by two polyvalent antivenins produced by Behringwerke in Germany. The ringhals is clearly a dangerous snake.

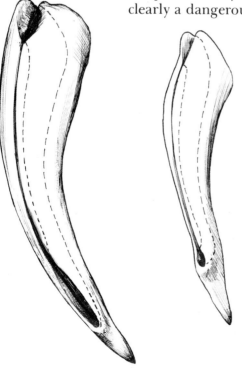

Two cobra fangs compared. The fang of a "normal" or non-spitting cobra at left has an elongated exit port toward the tip of the fang. The fang of the spitting cobra at right shows the smaller, rounder hole further up from the tip. Both holes, of course, are on the front or leading edge of the fang.

The pallid spitting cobra (*Naja* sp.) in the typical posture of a "spitter," mouth agape and eyes fixed on the target.

We said earlier that *spitting cobra* was a confusing name and indeed it is. The word *spitting*, to begin with, is inaccurate. In all cases the venom is *squirted* exactly as it is from a water pistol—by forcing a liquid out through a constricted orifice, properly aimed, at considerable pressure. Spitting is a word more properly reserved for what the hissing cobras do. When angry they expel air violently against a special bone in their glottis creating a very intimidating sound. As I can testify, it is explosive and startling. However, all of this notwithstanding, we will refer to the act of *spitting* venom simply because it is in such common use that any change attempted here beyond this explanation would be confusing and really academic.

It is generally stated that spitting cobras can spit their venom about twice their own body length with accuracy. That may be an exaggeration since the ringhals, a snake that seldom grows to be as much as five feet, can spit its venom to a distance of five to seven feet and is the most developed spitter of them all.

There is a deliberateness to the cobra's spitting technique. The head is held very still, the jaws part slightly, and twin jets of fluid are sent forth. The jets apparently break up into a mist before making contact as tiny droplets. The ringhals can spit several times in quick succession. (The ringhals is often encountered in popular literature under the name ring-necked cobra; to the Boers it is *spuw slang*.)

As with the other cobras the entrance point for the venom coming from the venom glands is at the top front edge of the tooth. (There is a duct for each fang, of course.) The emission port is also in front. But there the similarities end. In the non-spitters the exit hole is near the tip and is sloped downward just as in a hypodermic needle. The hole is characteristically ovoid and fairly large to accommodate a good flow of venom. In the spitting snakes the hole is back up from the tip and is both roundish and small. Inside the channel there is a sharp elbow bend that sends the venom out at right angles. A sharp contraction of the muscles surrounding the venom glands behind the eyes is enough to send the liquid through the ducts, into the fang through the entrance lumen, down the channel to the elbow bend, and out the small hole with considerable force.

In checking the popular literature one may find the suggestion that every species of cobra in Africa is a spitter. Without doubt this can be traced back to either nonsense tales or cases of mistaken identity in the field. There is no reason to think that any cobra in Africa is a true spitter other than the ringhals and the spitting cobra, *Naja nigricollis*. For all the rest, one must simply allow for such stories when hearing "tales out of Africa."

There seems to be general agreement that the spitting adaptation in the two African spitters and the one form of *Naja naja* found in Malaya is defensive. It is probably not used in food-getting because it is not needed. Why it should be limited to just three forms in a whole series of similar animals is not known. The ability is not identified with any viper probably because while cobra venom is able to quickly disperse through the thin membranes of the eye viper venom would be unable to due to its grosser molecular structure. At least this is one suggestion as to why the spitting ability is so limited. Certainly it is one of the more bizarre adaptations found in the adaptable world of snakes.

Su and his colleagues (1967) offer a concise and useful description of cobra venom: "Cobra venom is known to have a great number of pharmacological and enzymatic activities, reflecting the highly heterogeneous nature of its composition. The more important activities include the so-called neurotoxic, cardiotoxic, hemolytic, and histamine release. The peripheral neuromuscular blocking action, which leads to respiratory failure, constitutes the primary cause of death from the cobra venom in many animal species."

In any evaluation of toxic materials the sensitivities of test animals must be taken into account. What may be a swift lethal fraction for one test species may offer peripheral distress or at least embarrassment with slow onset for another. In the case of the cobra, however, respiratory failure seems to be a universal indication of impending mortality. In a sense, although using chemical rather than mechanical means, cobras suffocate their prey no less surely than pythons do.

NAJA COBRAS

The *Naja* cobras of Africa, the last of that continent's elapids, number four full species. Although there is a subspecies of the Asian cobra (*Naja naja oxiana*) found as far west as the Near and Middle East, we will limit our discussion to the full species of this genus that are clearly African animals.

Naja haje, the Egyptian cobra, is a five to eight foot animal found in the northern three quarters of the continent. It does not live in rain forests, however. Its venom is apparently about as toxic as that of the Asian cobra and although the animal is not particularly aggressive it is certainly a snake to be reckoned with. Antivenin is produced by Institut Pasteur in Paris and Behringwerke in Marburg, Germany.

N. melanoleuca, the forest cobra, is a six to eight foot snake that is generally restricted to areas of substantial timber. It is not any more aggressive than other cobras, but its venom is highly toxic and reportedly very fast-acting. Antivenins for this snake's bite are also produced in Paris and Marburg.

N. nigricollis is the fabled spitting cobra. Although not quite as well developed a spitter as the ringhals, it is clearly one of those few snakes able to discharge its venom over a distance of some

The spitting or black-necked cobra (*Naja nigricollis*) is one of the most accomplished of the venom squirters, second only to the ringhals. Its accuracy and range are usually exaggerated, however, particularly in popular accounts.

feet in defense of itself. It grows to be seven feet long and is said to be able to discharge its venom with some accuracy for nine feet. One of the most common snakes of the open grasslands, it is also one of the most dangerous snakes in Africa. For reasons not at all understood specimens of this species from northern areas produce a more potent venom than those from southern areas. The venom creates a marked increase in the histamine level of the blood, producing a shock effect. It acts as an anticoagulant at least at some levels of concentration. Much work remains to be done in studying the venom. Antivenins are produced in Paris, Marburg, and at the South African Institute for Medical Research.

The most toxic cobra venom belongs to *Naja nivea*, the seven-foot Cape or yellow cobra. It is not a particularly aggressive snake, but it is quick to display. When given the opportunity it will retreat. When it does bite, however, the yellow cobra causes a high mortality rate (high for cobras) and medical attention must be quick in coming if it is to have any effect. An antivenin to counteract the bite of this dangerous animal is produced by the South African Institute for Medical Research.

With the cobras of the genus *Naja* we conclude our discussion of the Elapidae. As a group they have been responsible for the death of uncounted numbers of human beings. Their diversity, their enormous range, their often formidable size, their extraordinarily powerful venoms, their sometime aggressiveness, and their great numbers make them a major public health consideration over most of the tropical and much of the temperate world exclusive of Europe. At the same time they are among the most interesting of animals. Their habits are largely unknown to us, the true nature of their venoms only barely understood, but better understood than most animal venoms, however. As a race we have worshipped them, adored them, decorated ourselves, our fantasies and/or symbology with their real and imagined powers. Many of them have learned to live in close proximity to man (in Bombay, India, a female cobra regularly nested behind books on the public library shelves) and they even serve us, for they kill tens of thousands of tons of rodents each year.

The elapid snakes are a fact of life in some of the most densely populated areas on earth. More human beings take them as an everyday event than do not, a fact that is hard to understand, perhaps, in our highly industrialized West. Our understanding of these strange and powerful creatures would seem to be a worthwhile goal.

The forest or black-and-white lipped cobra (*Naja melanoleula*) is a rather heavy terrestrial cobra that is not known to be particularly aggressive. Accidents with them are rare but some deaths have occurred. Lengths to over eight feet have been reported.

Chapter *13*

THE OLD

WORLD
VIPERS

The tic-palonga or Russell's viper
(*Vipera russelii*) is a large and statistically important snake in parts of Asia.
It was the "speckled band" of the Sherlock Holmes story. It is very prolific and
very common in some areas.

he family Viperidae, although widespread, is noticeably missing over vast portions of the globe. There are no representatives at all in either the Western Hemisphere, or the Australian-New Guinea-Solomon Island areas, however the rest of the tropical and temperate world is well supplied. With the exception of a single species of the family Crotalidae in the European region of the Soviet Union, the Viperidae account for all of the venomous snakes found in Europe and are all of the one genus *Vipera*. There are seven genera in Africa (including *Vipera)* and eight in Asia (again including *Vipera.*) All told there are ten genera of these highly advanced, often highly toxic, and occasionally very aggressive snakes.

The vipers are solenoglyphs, possessors of the most highly evolved type of fang mechanism. Unlike the proteroglyphs, the sea snakes and élapids, the vipers are not restricted in fang length by the capacity of the mouth. When not in use the viper's fangs are folded away in sheaths in the roof of the animal's mouth. Only when they are needed are they swung down into battle position and used as stabbing instruments. Coupled with a powerful toxin, the mobile, elongated, and easily replaced fangs of the vipers are terrible weapons. More than that, however, they are highly efficient food-getting tools.

Let us look at the Viperidae of Asia first, then those of Africa and finally Europe. The genus *Vipera* is the most widespread in the family. It is found in every area where vipers occur, with the single exception of southern Africa, and is often the only genus of Viperidae in evidence at all.

VIPERS OF ASIA

In western Indonesia, as far east as the Viperidae can be found, there ranges a single species of the genus *Vipera, V. russeli,* Russell's viper or the dreaded tic-palonga. This is the "speckled band" of Sherlock Holmes fame and is one of the most seriously venomous snakes on earth. The species has an enormous range and is found in India, Pakistan, Sri Lanka, Burma, Thailand, China, and Formosa. It ranges from sea level up to at least 7,000 feet. On the mainland, specimens over five feet are not uncommon but on offshore islands they tend to be smaller. The tic-palonga lives on small mammals, lizards, birds, and amphibians. It has the unfortunate habit of lingering on footpaths and roadways and even entering human habitations in search of food. Bites, therefore, are common. It is an extraordinarily prolific species giving birth to as many as 60 living young at a time. The newborn may be almost a foot in length and can be troublesome from the outset.

The tic-palonga is not a fast-moving snake until it strikes, but then it becomes exceedingly agile. It is mainly nocturnal and although apparently reluctant to bite it can strike with enormous force and determination.

Bites by the old-world vipers (also called, and properly so, the true adders) are generally slower to kill than serious elapid intoxications. However, in at least one instance, a bite by a Russell's viper led to a very quick death. The medical records for 1921 in Ceylon (now Sri Lanka) contain the case of an adult woman bitten by a Russell's viper said to be about five feet long. It evidently struck a vein, something of a rarity in snakebites, and she died within 15 minutes. Elapid venoms seem to come in smaller molecules and are probably able to disperse more rapidly than viper venom and thus act more quickly. That would be a general rule, but exceptions, as we have noted, are not unknown. At any one time an adult Russell's viper is probably capable of discharging between 150 and 250 milligrams of venom. It probably would take less than 70 milligrams to kill even a very large man. Fortunately, a number of antivenins are produced by Behringwerke (Germany), Central Research Institute and Haffkine Institute (India), and Queen Saovabha Memorial Institute (Thailand.)

In Eastern and Southeastern Asia (an area that includes Tibet, China, Korea, Japan, Burma, Thailand, Laos, Cambodia, and Malaysia), we find the Viperidae represented by two genera—*Vipera* and *Azemiops.* The *Vipera* is again *russeli* only somewhat larger than the members of this genus found on the islands. But, the snake's habits and the problems stemming from them are the same. It is a large and highly toxic snake which lives near human settlements and even enters human habitations at night. The other true viper is *Azemiops feae,* a small animal, and apparently a very rare one, seldom over three feet long, known as Fea's viper. Very little is known about its venom or its effects on human beings. It is assumed that under the appropriate conditions it could be dangerous to a human being, but there are no data to substantiate this assumption. No antivenin is available.

Once we reach Central Asia in our march westward (an area we will designate as Afghanistan, Pakistan, India, Sri Lanka, Kashmir, Nepal), we find four genera of Viperidae in residence: *Echis, Eristicophis, Pseudocerastes,* and *Vipera.* The first three are represented by one species each, the last by two.

The genus *Pseudocerastes* is represented by *P. persicus,* the Persian horned viper. This nocturnal, three foot long snake is known to be dangerous and can be both active and aggressive once night falls. Bites are less likely during the daylight hours when the snake tends to be sluggish. An antivenin is produced by the State Razi Institute (Institut d'Etat Des Serums et Vaccins Razi, Teheran, Iran).

The genus *Eristicophis* is represented by the Asian sand viper (*E. macmahonii*). Although less than three feet long, this arid land snake can be dangerous. No antivenin is produced for its bite.

The two representatives of the genus *Vipera* are distinctly dangerous. *V. russeli,* Russell's viper, again appears and is no less dangerous in Central Asia than in Eastern or Southeastern Asia.

The Asian sand viper (*Eristicophis macmahonii*) comes from Iran, Afghanistan, and Pakistan and, although a small snake (under three feet), it may be considerably more dangerous than previously thought. It is a species requiring further study.

The second member of the genus, *V. lebetina*, the Levantine viper, is a highly toxic five-foot animal that although placid in daytime is alert and occasionally savage at night. The quantity of venom it delivers (and its toxicity) is believed to be about the same as that of Russell's viper. This snake is so placid in the daytime that its various native names translate roughly as *blind one* or *deaf one*. The latter name is, of course, quite accurate. Antivenins for the Levantine viper are produced by the Institut Pasteur in Paris, the Razi in Teheran, and the Tashkent in Moscow.

It is when we come to the genus *Echis* that we encounter a snake of special significance. *Echis carinatus*, the saw-scaled or carpet viper, is one of the most dangerous snakes in the world from the point of view of health statistics. In many areas it is *the* major cause of snakebite. While not a large snake, generally under two feet long, it is short-tempered, even aggressive, and very toxic. Its range is enormous, stretching from Sri Lanka and India, across all of western Asia and northern Africa, and down into tropical Africa—wherever there are dry zones. In eastern Egypt, Israel, and the Arabian Peninsula it is replaced by the apparently equally dangerous *Echis coloratus*, also called the saw-scaled viper. (An antivenin is produced for *E. coloratus* by Rogoff Medical Institute in Israel.)

Pope's pit viper, *Trimeresurus popeorum,* a common danger to workers on tea estates between 3,000 and 5,000 feet above sea level in Assam, Burma, Cambodia, Malaysia, and Indonesia. It is usually not lethal but decidedly dangerous, one of about 30 species of generally similar snakes found in eastern Asia.

It has been postulated that 8 milligrams of *Echis* venom would be enough to kill a man. In calculating minimum lethal doses for man, Chaudhuri and his co-workers (1971) created the following comparisons: *Echis* venom required to kill an adult human being need be only one-third the quantity of common cobra (*N. naja*) venom needed to accomplish the same end; one-half the amount of banded krait (*B. fasciatus*) venom, less than one-eighth of the Russell's viper (*V. russeli*) venom, and one-twentieth of the green pit viper (*Trimeresurus gramineus*) venom.

The reports of mortality rates from *Echis* bites are confusing. Ahuja and Singh (1956) report an Indian series of 58 bites with 19 deaths, a mortality rate of 32.8 percent, a very high mortality rate for any venomous animal. Yet, in a series reported from the Sudan (Corkill, 1956) there were 21 cases with *no* deaths. An African series reported by Chapman (1968) included 19 *Echis* bite cases with a mortality rate of 20 percent. Reid (1968) reported on death times for three cases; they ranged from 27 hours to an extraordinary 41 days.

In many areas the saw-scaled viper is very common. In an interesting historical note, Deoras (1971) reported a period in Indian history when a bounty was paid for anyone killing and presenting the body of a saw-scaled viper. In 1890, in Ratnagiri District of Maharashtra State, 225,721 vipers were presented for bounty. In the same year, in the same district, that species killed 123 people. One wonders how many of those victims were bitten while attempting to capture or kill snakes for money.

The list of active proteins in *Echis* venom is long, complex, and certainly not fully understood. While neurotoxic factors are present, they are probably not of clinical importance. There are hemorrhagic factors that are very significant and cause severe bleeding from the gums and the nose, and in the kidneys. In addition, there are proteolytic enzymes and factors that modify coagulation of the blood, including both accelerators and inhibitors. Besides severe bleeding (which can be fatal) there are fever, pain, and swelling at the site of the wound. Fortunately, there is antivenin available in several polyvalent as well as monovalent forms. At least five of the antivenins produced by the Institut Pasteur in Paris are specific to this snake. Antivenins are also prepared by the Tashkent in Moscow, the Central Research Institute and Haffkine Institute in India, and the Institut d'Etat Des Serums et Vaccins Razi in Teheran.

Quantification of the danger presented by the saw-scaled viper cannot realistically be attempted. We do not know how many people are bitten over this snake's vast range, we do not know the margin that should be allowed for error in field identification, and we do not know the mortality rate among the victims. Reports and conversations with field workers do seem to indicate, however, that *Echis carinatus* and perhaps its congener, *E. coloratus,* are among the most troublesome venomous creatures on earth.

In Western Asia (an area including Turkey, Cyprus, Asiatic Egypt, Israel, Jordan, Lebanon, Syria, Saudi Arabia, Yemen, Handramat, Iraq, Iran, and Transcaucasian U.S.S.R.—after Henriques and Henriques, 1971) there are five genera of old-world vipers. One species of the genus *Bitis* occurs, *B. arietans* the puff adder. We will examine this snake in detail when we discuss Africa, where *Bitis* is a genus to be reckoned with. *Echis,* as we have noted, is represented by the two saw-scaled vipers. *Pseudocerastes persicus,* the Persian horned viper, another snake we have made reference to, also occurs.

The genus *Cerastes* is represented by *C. cerastes,* the desert horned viper. Although not a particularly large or dangerous snake, it is disagreeable and very quick to strike. Bites are common in many areas and antivenins are prepared in Algeria, France, Germany, and Iran, giving some indication of the concern this snake causes. (*C. vipera,* the Sahara sand viper, is also found in a few areas.)

Atractaspis is represented by two Middle East mole vipers—*A. engaddensis* and *A. microlepidota.* The former ranges in Egypt and Israel and the latter on the Arabian Peninsula as well as over much of Africa. These snakes are under three feet long, but they have large fangs. They are nocturnal and are often found around oases, places where people and livestock are likely to be. Their bite typically causes local pain and swelling and there have been some deaths reported. In severe cases there will be fever, vomiting, and blood in the urine. No antivenin is produced to counteract the bite of these troublesome animals.

There are five members of the genus *Vipera* in Western Asia. *Vipera lebetina* and *V. russeli* we have already examined in some detail. In addition, there are *V. ammodytes,* the long-nosed sand viper (antivenin prepared in Yugoslavia, France, Germany, Iran, and Italy) and the relatively innocuous European viper, *V. berus.* Although small and not particularly dangerous this latter viper is incriminated often enough in bites on humans to warrant the production of antivenin in France and Italy.

The Near East viper, *V. xanthina,* is a snake with a short temper. It is quick to strike and although mortality is reportedly about 5 percent among humans it does exhibit a familiar and unfortunate set of habits—it is nocturnal and fond of living near man. An antivenin is produced by Institut Pasteur in Paris.

VIPERS OF AFRICA

In Africa, we find both larger snakes and much more dramatic forms than in Asia. Vipers of North Africa include genera *Bitis, Cerastes, Echis,* and *Vipera.* As we move down into Central Africa

The business end of a puff adder (*Bitis arietans*). The fangs can be seen still folded back in their sheaths. This is a female specimen and her newborn young can be seen nearby.

we encounter genera new to this discussion—*Atheris, Causus,* and *Adenorhinos* plus five genera we have already discussed. Start by the time we reach the southern third of the vast African land mass, we will see Viperidae whittled down to three genera we will already have explored. It is in tropical central areas of Africa that the Viperidae come into their own—eight genera represented by scores of species and subspecies, including some especially colorful and dangerous animals and, without a doubt, the largest of all the vipers.

Central Africa is a vast area. From the Sahara's southern edge it reaches southward until well after the continent has necked down into a virtual peninsula running toward the Cape Province in South Africa. Between 30 and 35 political units are included within the general region and a vast number of tribes, languages, dialects, customs, and levels of cultural attainment.

Through this incredible mélange thread the enormous vipers of Central Africa, huge, toxic, and often plentiful. One, at least, has the longest known fangs of any snake, larger by far than those of even the king cobra, the black mamba, or the dreaded Australian taipan. (Those three elapids are probably the longest venomous snakes in that order. To recapitulate observations previously made, maximum lengths for the three giants are usually given: king cobra—close to 19 feet, black mamba—14 feet, taipan—11 feet. It is possible that a member of the Crotolidae, the bushmaster, may reach 12 feet in extreme cases.)

The Vipers of Central Africa

Genera	Common Group Name
Atheris	African bush vipers
Atractaspis	mole vipers
Bitis	African vipers
Causus	night adders
Cerastes	horned vipers
Echis	saw-scaled vipers
Vipera	True adders
Adenorhinos	worm-eating viper

Of these eight genera several are of no great consequence in the public health files. The newly named genus *Adenorhinos* with its single species of worm-eating viper, *A. barbouri,* is an example. This viper is small, retiring, and although venomous not seriously so. It is certainly possible for someone handling a specimen to get bitten, but local pain is all that might be expected. Of course, as with any injury, particularly involving puncture wounds, serious infections and tissue damage might ensue. No antivenin is available for *A. barbouri.*

Atheris is another genus of little serious consequence. The African bush vipers are small (all under three feet) and their bites are nowhere near as serious as those of many of the other vipers of Africa. *A. nitschei,* the sedge viper, *A. squamigera,* the green bush viper, and *A. chloroechis,* the West African bush viper are examples of these small, prehensile-tailed, and largely arboreal animals. No antivenins are prepared for their bites and very little is known about their venoms. Several other species of *Atheris* occur in addition to those named.

There are a dozen or more species of mole vipers, genus *Atractaspis,* a genus we met in the Middle East. *A. bibronii* and *A. corpulenta,* Bibron's mole viper and the western mole viper respectively, are examples. These animals are short-tempered and quick to bite. They are notoriously difficult to handle and snake handlers attempting to capture specimens are subject to injury. Their bites are painful but for the genus as a whole not usually very dangerous. Mole viper bites may cause local inflammation and hemorrhage, restlessness and weakness, but no paralysis. Breathing may be difficult, but is usually not seriously impaired. There is a notable exception, however. *A. microlepidota,* the northern mole viper, may have a mortality rate among its victims as high as 20 percent. That is extraordinarily high and would seem to warrant further investigation. The genus is reputed to have enormously enlarged fangs. This is only partially true—the fangs are very large for these snakes' notably small heads, but when compared to other vipers are not especially outsized. Although some members of this genus are highly toxic there is no antivenin produced for their bites. It is unusual for a snake to be able to deliver bites 20 percent of which are fatal to man without an antivenin being available. This is especially so since *A. microlepidota,* possibly the most dangerous member of the genus, is a common snake and bites are not unusual. That 20 percent figure is obviously in need of further review.

Causus, the night adders, are not uncommon in many areas of Central Africa. There are four species and all are under three feet in length. *C. rhombeatus,* the rhombic night adder, is perhaps the best known. They are nocturnal, as their name implies, and have small fangs. Their venom is only mildly toxic to man, but it is painful locally. Severe systemic reactions do not occur, usually, and no antivenin is produced. Snakes of the genus *Causus* are generally regarded as the most primitive of all the Viperidae.

We again encounter the saw-scaled viper, *Echis carinatus,* in Central Africa and it is no less a problem here than in the other regions where it ranges. It is a ferociously short-tempered snake that bites first and "asks questions afterwards." It is not a snake to step on wherever it is found, the more so if you are a barefoot indigenous person.

The widespread Asian, Middle Eastern, and African species, the saw-scaled viper (*Echis carinatus*) is one of the most troublesome venomous snakes on earth. It is short-tempered, often quite common and widespread in dry areas with large barefoot populations. It *might* be the single most statistically dangerous species in the world.

The African desert horned viper (*Cerastes cerastes*) is not a particularly dangerous snake although some sub-lethal accidents do occur. Like all true desert snakes it is nocturnal and can make itself all but totally invisible by wriggling down into the loose sand. Not all specimens have such well-pronounced "horns."

Cerastes crosses our path again—*C. cerastes*, the African desert horned viper, and *C. vipera,* the Sahara sand viper. Both, as their names imply, appear just in the arid margins of the central African region. They are not very dangerous although they can be short-tempered. Antivenins for both snakes are produced in both France and Germany. In dry zones these are animals to be regarded with care, but for the African continent overall their consequence is slight.

The true adders, *Vipera,* appear again although in the region we are discussing they are of minor concern. *V. superciliaris,* the African lowland viper, is rare and there are no bites on record for this animal. *V. mauretanica,* the Sahara rock viper, has been the cause of some isolated cases of distress and an antivenin for its bite is produced by the Institut Pasteur in Algiers. *Vipera* is not a major cause for concern in Central Africa.

The trouble starts when we come to the genus *Bitis.* These are the largest and most heavy-bodied vipers in the world and some, at least, have the longest fangs of all venomous snakes. One species kills more Africans than any other snake on the continent (and when you take into account the mambas and cobras, that statement is worthy of careful consideration) and few species within the genus may be considered trivial in their public health consequence. *Bitis* is the genus within which the vipers of Africa come into their own. Because of their size, their extreme toxicity, and their often spectacular markings the *Bitis* of Central Africa are among the most dramatic snakes on earth.

For a number of reasons, clarity being the foremost, we will examine genus *Bitis* for the entire continent south of the Sahara, considering Central Africa and Southern Africa together:

The Genus BITIS South of the Sahara

Species	Common Name (s)	Maximum Length
B. arietans	puff adder	5'
B. atropos	Berg adder, Cape mountain adder	15"—20"
B. caudalis	horned adder, Cape horned adder, many-horned adder	12"—13"
B. cornuta	Hornsman adder	12"—15"
B. gabonica	gaboon viper	6½' on record
B. heraldica	Angola viper, Bocage's viper	
B. inornata	Cape puff adder	
B. nascicornis	river jack, nose-horned viper, rhinoceros viper or adder	4'
B. paucisquamata	spotted dwarf adder	Under 12"
B. peringueyi	Namib desert viper, side-winding adder, dwarf sand adder, Peringuey's desert adder	10"
B. worthingtoni	Kenya horned viper, Worthington's adder	

By no means are all of these species of equal interest or concern. *B. atropos*, an irascible little snake under two feet long, is of interest because its venom appears to be largely neurotoxic and even quite elapid-like. The effects of *B. cornuta*, Hornsman adder, venom have apparently been exaggerated. Stories about the Hornsman that had some considerable currency were based on indigenous legends rather than on clinical reports. Indigenous legends are about the last place one turns for information about toxicity.

The horned puff adder or South African horned viper, *Bitis caudalis,* is a small snake usually well under a foot and a half in length. It is, however, highly toxic and some human deaths have been attributed. It hides in the sand and strikes out without seeming provocation.

The gaboon viper (*Bitis gabonica*) is one of the most colorfully marked snakes in the world and is the species with the longest fangs known. It is a seriously venomous African species with an enormously heavy body and large head.

Too little is known about *B. heraldica*, the rare *B. inornata*, or *B. worthingtoni* to allow for discussion. It should be noted in passing, however, that although the common name for *B. inornata* is Cape puff adder there is no further comparison to be made. The African puff adder, *B. arietans*, as we shall be seeing, is quite a different kind of beast.

It is when we discuss two species from this genus that we can understand the great interest that has been shown in the group. *Bitis arietans* and *Bitis gabonica*, the African puff adder and the Gaboon viper, are formidable animals.

B. gabonica, the Gaboon viper, grows to be well over six feet long. It is enormously heavy-bodied with a head that seems the size of a small plate. It is the longest and heaviest snake of its group, as we have said, and it carries the longest fangs known of any snake, nearly two inches in a very large specimen. The bite of this snake resembles the typical viper bite we will be discussing shortly and elapid envenomation as well. The bite causes marked neurotoxic effects as well as great and damaging local involvement. An enormous dose of venom can be delivered by a really large gaboon viper, certainly by one six feet long, and the enormous fangs can deliver it to areas deep in well-blooded tissue. A barefooted native struck on the foot or lower leg by a snake with two-inch fangs would be very nearly bitten through. No other snake is capable of reaching the tissues available to a large gaboon viper. Effects reported from the bite of this species include extensive and highly damaging extravasation (the forcing of blood out of the blood vessels into surrounding tissue), depressed heart action, blood in the urine, and extreme dyspnea or difficulty in breathing. Because of the great quantity of blood that may be forced into the tissues around the bite there may be permanent damage to a limb. Necroses may be extensive and very difficult to deal with clinically.

In January, 1964, the zoo world was saddened by the death of Jerry de Bary, the thirty-seven-year-old director of Hogle Zoological Garden in Salt Lake City. De Bary was working late on a Saturday evening and decided to clean the cage of one of his prize specimens, a five-foot African puff adder (*B. arietans*). Apparently, he suffered a slight dizzy spell just as he opened the snake's locked cage and was off guard long enough to be struck on the left forearm. He managed to close the cage door but then collapsed. The night watchman found him and called his wife.

Taken to the hospital in a police ambulance, de Bary revealed that the zoo did not have a supply of antivenin. This, of course, was a mistake, a fatal one as it turned out. A Navy jet flew a limited supply of the proper serum from the zoo in San Diego, California. It was not enough to treat the case. De Bary was reportedly in agony and his heart stopped several times. A resuscitator and heart stimulants were used, and a tracheotomy was performed. The zoo man died while additional antivenin was en route by air from South Africa.

Three and a half decades earlier another zoo man, Marlin Perkins, was bitten by a gaboon viper. He described the pain as extensive and intense. "I felt like I had a big weight on my chest and I wasn't able to breathe. And it felt like acid was eating inside of me," Perkins said.

The puff adder (*B. arietans*) is the second longest of the vipers, but it has a number of other superlatives all to itself. Often five feet long and grossly heavy in the body (with an absolutely enormous head in large specimens), the puff adder probably kills more people in Africa than any other snake. *It causes more cases of snakebite than all of the other venomous African snakes combined.* If it is second to the gaboon viper in length, it is second only to the black mamba in the fear it instills. It is one of the most common of African venomous snakes and has one of the widest distributions. Puff adders are often found heavily infesting an area and can cause considerable difficulty for the human population. The snake feeds, by and large, on warmblooded prey and often seeks rodents around human habitations. It is not at all shy about entering a home at night, for it is largely nocturnal, and has a reputation for being placid that can be misleading. This reputation apparently derives from the fact that the snake is usually only observed during the day when it is admittedly less active and the fact that observers clearly mistake one of the snake's worst characteristics for one of its better points. The puff adder is perhaps more stubborn than placid. It is reluctant to move even when disturbed, move away that is. Even when it senses someone or something coming down the trail it will hold its ground rather than flee as so many other snakes do almost automatically. This behavior results, of course, in a great many very large puff adders being stepped on by a great many barefooted people in the process of becoming statistics. It is undoubtedly this immobile, nocturnal pattern of behavior when coupled with the widespread occurrence of the species that accounts for the high incidence of bites.

The toxin delivered in a puff adder's bite is very potent and is frequently given in large quantities. It has been postulated that enough is presented in a single bite to kill four or perhaps five large men. The bite is always characterized by enormous swelling. The onset can be rapid and even a victim bitten on a single toe may suffer gross swelling all the way to the groin. There is pain, a great deal of it, massive shock, and extensive tissue damage. Convulsions, fever, and unconsciousness may all appear as symptoms. Great necrotic areas may develop, but probably not due directly to the venom. As is the case with the gaboon viper, extravasation is apparently the cause.

Puff adder victims suffer from a peculiar secondary syndrome. Hours after the patient seems past the critical point, when they seem almost certainly to be getting better, there will be a sudden collapse and possible death. This occurs most often in patients exhibiting extensive extravasation. They suffer from the kind of symptoms seen in traumatic injuries of the crush type. Death may be due in these cases to loss of blood through bleeding into the tissues or from other causes not really very well understood. Such patients are probably best placed in the hands of physicians specializing in trauma.

Visser (1966) presents some interesting case histories of South African venomation. One caused by a puff adder is adapted here: CASE—24-inch puff adder, single fang bite on left thumb knuckle.

0 minutes	No symptoms recorded.
10 "	Sharp pain area of bite.
20 "	Thin watery blood oozing from incisions.
40 "	Thumb turns blue.
45 "	Throbbing pain at site of bite.
55 "	Pain intensifies, swelling progresses to forearm.
2:00 hours	Victim stands up then passes out.
2:30 "	Regains consciousness; thirsty and hungry. Coagulated blood at incision dislodged by throbbing.
3:00 "	Dizziness returns.
6:15 "	Feels weak. Walk to camp dispels dizziness.
6:25 "	Swelling increases beyond elbow.
7:30 "	Pain acute, progressing to shoulder. Craves liquid.
9:00 "	Sleeps four and a half hours. Vomiting, pain prevents further sleep.
18:00 "	Throbbing pain replaced by burning sensation.
24:00 "	Large blisters in area of bite.
27:00 "	Pain diminished; swelling unchanged, arm itchy.
4 days	Dizzy spells persist; thumb greenish yellow. Swollen glands.
8 days	Stiffness of knee joints.

Thumb sloughed periodically and remained immovable for four weeks. After three months thumb still not functioning normally and patient was still prone to sudden fainting spells or blackouts.

In one series of puff adder bites (1957-63) Chapman (1968) reported 210 cases. There were 56 cases with severe reactions and 11 deaths, yielding a mortality rate of 5.2 percent. Elsewhere, Corkill (1956) suggests a 7 percent mortality rate. Both figures are surprisingly low when one considers the size and nature of the animal under dicussion.

Bitis nascicornis, the river jack or rhinoceros viper, is probably at least as venomous as either the gaboon viper or the puff adder. It is also thought to be more active and more likely to strike out at things nearby. However, it is relatively rare and therefore not so often encountered clinically as the other two giants. The river jack grows to be over four feet long.

As might be expected for the snake that bites more people than all of the other African snakes combined and kills more people than any other single species, there is antivenin to help counteract its effects. The genus *Bitis* is fairly well covered, as will be seen in the following table:

ANTIVENIN FOR *BITIS*

Source	Product Name	Species Covered
Institut Pasteur (France)	Serum Antivenimeux *Bitis*	*B. arietans* *B. gabonica*
	Serum Antivenimeux *Bitis-Echis-Naja*	*B. arietans* *B. gabonica*
Behringwerke AG (Germany)	Serum Nordafrika	*B. gabonica* *B. arietans*
	Serum Zentralafrika	*B. gabonica* *B. arietans* *B. nasicornis*
South African Institute for Medical Research (South Africa)	Polyvalent Antivenin *Bitis-Naja-Hemachatus*	*B. arietans* *B. gabonica*
	Polyvalent Antivenin *Hemachatus-Naja-Bitis-Echis*	*B. arietans* *B. gabonica*
CAPS (Southern Rhodesia)	CAPS Snake Bite Antivenin	*B. arietans* *B. gabonica*

In passing, we should note that *Atractaspis* with at least two species and *Causus* with another two are also found in Southern Africa—along with, of course, *Bitis*. *Atractaspis* particularly are irascible and quick to bite. They have a peculiar method of doing so. Their lower jaw is underslung and when the fangs are lowered preparatory to an attack they protrude on either side of the bottom jaw. In the bite itself the head appears to be shot forward, or perhaps sideways, *above* the bite site and the fangs (often only one makes contact) are brought down sharply in a stabbing motion. There have been many reports of the snake holding on and even attempting to encircle the bitten member with its neck, apparently in the effort to pump extra venom into the wound. *Causus,* the night adders, are the same in South Africa as they are further north, generally retiring and inoffensive. Where they got their name *demon adder* is a mystery. They are toxic, of course, but not very seriously so and they are apparently reluctant to bite unless forced into it.

With our discussion of the Viperidae in Africa we conclude our herpetological references to that continent. Africa has been for so very long the epicenter of legend and tales of high adventure that the uninitiated still conjure up visions of the "darkest continent" fairly swarming with venomous snakes. This, of course, is not at all the case. As I can testify it is quite possible to travel for long distances in Africa without so much as seeing a single snake. On two trips my wife and I traveled overland for many hundreds of miles and saw only one road-killed cobra, no other sign of a snake. Admittedly our interests at that time were not herpetological and we were not overturning rotten logs and rocks in search of herpetofauna, but a land *crawling with snakes* would certainly have provided an occasional chance encounter.

Head of the spectacular riverjack or rhinoceros viper (*Bitis nasicornis*). This is an African species with a relatively limited range and bites are not often reported. Its venom is extremely toxic, however. It is a very heavy-bodied animal.

Snakes in Africa as elsewhere are generally retiring. Some few species, as we have seen, are active at night and are therefore easy *not* to see. Because these nocturnal species are either stubborn or lethargic (interpretation open and personal), they do tend to get involved with people in a rather unfortunate manner. Still, snakebite deaths are the exception and incidents themselves rare when one considers what the potential really is. It is safe to say that if snakes were even a bit more aggressive than they are the snakebite accident and death figures, whatever they may actually be, would soar. Clearly snakes like meeting man, if anything, less than man likes meeting snakes. It is not being facetious to observe that no snakes collect and exhibit men. And a great many of the accidents with venomous snakes are the direct results of men attempting to handle snakes for whatever their motives may be. With many species it is almost impossible to imagine an accident when the snake is not being purposefully handled. Records indicate that for a number of species all acidents have been caused in just that way.

VIPERS OF EUROPE

In all of Western Europe there are only seven species of venomous snakes and all are of the genus *Vipera*. A single species of the genus *Agkistrodon* (an important genus of the family Crotalidae that we will discuss next) extends into European U.S.S.R. from Asia. There is no other situation comparable to this, where an area the size of Western Europe is limited to a single genus of venomous snake. By comparison, the United States alone has five genera, Western Asia has nine, and Central Asia has ten. It is true, of course, that the tropics are much more richly endowed with reptiles in general, but Western Europe is not all Sweden and Finland. It includes warm Spain, Portugal, and Greece, too. It is just that Europe is poorly endowed with venomous forms (or, perhaps, richly endowed with a paucity of them).

The only venomous snake in northern Europe is the European viper, *Vipera berus*. It ranges across Europe and Asia all the way to northern Korea. In southern zones it generally stays to the mountains up to an altitude of 9,000 feet. Two feet is the general maximum length given for this snake although in one grandfather (actually, grandmother since in *V. berus* the females tend to be larger) a length of 34 inches was recorded.

Generally speaking the European viper is timid, but it will strike quickly if frightened or stepped on. Minton and Minton (1969) report that there were 163 snakebite cases in Finland alone in 1961, so apparently enough Europeans do not look where they are going to give us some statistics. Most injuries are to hikers and agricultural workers, and that is generally true across Europe even though *V. berus* is not uncommonly found in stone walls and around overgrown gardens. Its venom is capable of killing a man although that is clearly the exceptional case. It has been postulated that the venom is about one-fortieth as toxic as Russell's viper venom.

In western-southern Europe *Vipera aspis*, the asp viper, partially replaces *V. berus*. It is an 18 to 24 inch snake that is generally sluggish and frequently nocturnal. *V. aspis* is believed to be about as toxic as its northern cousin.

On the Iberian Peninsula and across into northwest Africa, ranges *V. latasti*, the snub-nosed viper. Not much is known about the species, but it is believed to have a fairly mild toxin. It is, however, distinctly venomous and of some potential danger to anyone sensitive to foreign proteins and enzymes.

There are a few other species: *V. kaznakovi* and *V. ursinii* range into southeastern Europe from Asia. The former is found in the Caucasus Mountains and in Turkey. Both *V. xanthina*, the Ottoman viper, and *V. lebetina*, the Levantine viper, are seriously venomous Middle Eastern species that reach Europe in Turkey and on some eastern Mediterranean islands. Their appearance in Europe is almost accidental.

It is generally thought that *Vipera ammodytes*, the strikingly beautiful three-foot long-nosed viper, is the most seriously venomous snake in Europe. It is found in southeastern Europe and in Asia Minor. *V. ammodytes* is quick to strike when disturbed and is apparently highly toxic, causing both severe local and possibly fatal systemic reactions.

Vipera berus, the European viper in common and melanic phases. This is a relatively mildly toxic snake with very few human deaths reported from its bite. It is unreasonably feared in some places.

The Levantine viper, *Vipera lebetina,* a species that grows to over five feet and is sometimes referred to as the giant Russian viper. It ranges from Cyprus through the Middle East and into European Russia. It is a dangerous snake, more so at night, but not to be trusted at any time. It is partially arboreal or at least is found in bushes.

The only other venomous snake in Europe is limited to an area between the Volga River and the Ural Mountains. It is *Agkistrodon halys*, Pallas's viper, a pit viper closely related to our copperhead and cottonmouth water moccasin. It is of a species, genus, and family we will be discussing in the next chapter.

The venomous snakes of Europe are less of a problem than their counterparts in any other part of the world. Still, to the individual victim, a venomous snakebite is bad without regard for geography. *V. aspis*, for example, can cause acute pain which will spread throughout the entire limb. (Virtually all snakebites are on extremities—fortunately. Head and torso bites are much more difficult to deal with both because a tourniquet or ligature is not possible and because the venom enters the body nearer to vital organs.) The pain is accompanied by swelling and discoloration. Some hemorraging occurs and can progress far enough to cause permanent tissue damage. Systemic reactions can include shock, vomiting, pulse irregularities, and a feeling of giddiness. Gangrene can ensue and there may be kidney disorders and even liver involvement. If the venom is introduced into a blood vessel, even a very small one, rapid death can follow. The snakebite problem in Europe may be relatively infrequent, but it can be serious.

Given the general sophistication of European society and social services it is not surprising that the continent's venomous snakes are well covered by antivenins. The following are currently available:

The mamushi or Pallas's viper (*Agkistrodon halys*) is found in Japan, Korea, and China although it is related to the cottonmouth water moccasin and the copperhead of North America. It is a shy snake but crowded human conditions within its range result in from 2,000 to 3,000 bite cases a year in Japan alone. Few if any are fatal. The snake is used for food and medicine and is handled in the marketplaces as if it were harmless.

Source	Product Name	Species Covered
Institut Pasteur (Algeria)	Serum Antiviperin	*V. lebetina*
Institut Pasteur (France)	Serum Antivenimeux *Berus-Ammodytes*	*V. berus* *V. ammodytes* *V. aspis*
	Serum Antivenimeux *Lebetina-Xanthina*	*V. lebetina* *V. xanthina*
	Serum Antivenimeux *Aspis-Berus*	*V. aspis* *V. berus*
Behringwerke AG (Germany)	Serum Nordafrika	*V. lebetina*
	Serum Naher—and Mittlerer Orient	*V. ammodytes* *V. lebetina*
	Serum Europa	*V. ammodytes* *V. lebetina*
Institut d'Etat Des Serums et Vaccins Razi (Iran)	Polyvalent Antivenin for Iran	*V. lebetina* *V. xanthina* *V. ammodytes* *Agkistrodon halys*
	Polyvalent Antivenin for Middle East, India, and Pakistan	*V. lebetina*
Instituto Sieroterapico e Vaccinogen (Italy)	Antiophidio Serum	*V. ammodytes* *V. aspis* *V. berus*
Tashkent Institute (Russia)	Gyurza Antivenin	*V. lebetina*
	Gyurza Polyvalent Antivenin	*V. lebetina*
Institute for Immunology (Yugoslavia)	Serum Antiveperinum	*V. ammodytes*

Chapter 14

THE

PIT
VIPERS

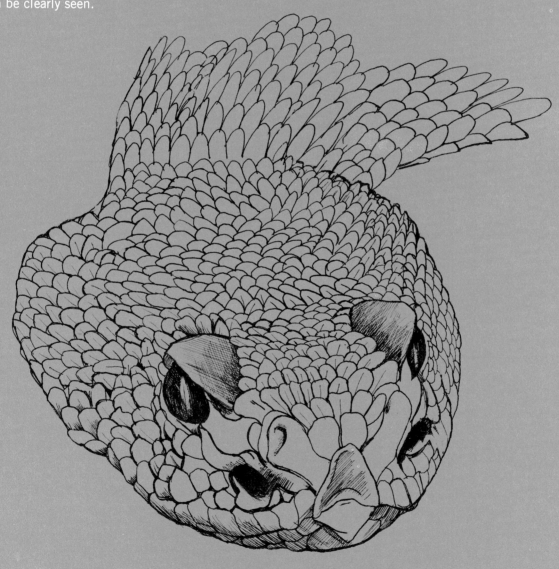

Head-on view of a composite pit
viper. The forward orientation of the
pits can be clearly seen.

he last group of snakes we will explore are members of the important family Crotalidae—the pit vipers. The family reaches maximum diversification in the Western Hemisphere; there are five genera in North, Central, and South America while all of Asia westward to European Russia has only three. There are no members of this family in Africa or in the Australia-New Guinea-Solomon Island region.

Like the Viperidae, the Crotalidae are solenoglyphs. They have rotating fangs that can be stored in the roof of the snake's mouth when not in use which, again, allows for longer fangs. This is the most advanced type of fang system. It is often suggested that the Crotalidae descended from the Viperidae (or from a common ancestor) and represent a kind of ultimate refinement. (It will be noted here that we are aligning ourselves with those who feel the pit vipers are properly placed in a family of their own rather than in a subfamily of the Viperidae. The crotalids' fang system, however, is a duplicate of that found in the old-world vipers.)

The immediately visible feature that separates the Viperidae and the Crotalidae is a heat-sensing pit. In all members of the family the pit appears between the eye and the nostril, rather nearer the latter. If you draw a line between the nostril and the eye (through the middle of each), the pit lies slightly below the line.

It was once thought that the pit was for sensing movement, for enhanced smelling, even for hearing, but in recent times it was shown that its purpose is heat detection. Up to a distance of about a foot the snake can detect radiating body heat of prey animals. The aperture points forward, as might be expected, and is divided into two internal chambers. They are separated by a membrance and the after chamber has an air duct that runs back and up to the snake's eye socket which allows equalization of air pressure on either side of the delicate membrane. The incorporation of this device into the snake's face required an evolutionary modification of the upper jaw, the jaw that carries the rotating fangs and along which the venom ducts run. There is a dent in the upper jaw of the Crotalidae not found in the Viperidae.

Very often, when we think of the new-world pit vipers, we think of the rattlesnake's rattle. This strange device is limited to two genera—*Sistrurus* and *Crotalus*. It is generally believed that all species within these two genera have rattles and snakes not in them do not. (There is one rattleless rattlesnake, *C. catalinensis*, the Santa Catalina Island rattlesnake.) It is interesting to note that other crotalid snakes, animals in the genus *Bothrops* and *Agkistrodon* and perhaps others as well, tend to move their tails in a nervous twitching action under the same circumstances that would inspire a rattlesnake to use its tail. In fact when this is done in dry leaves, the rattleless crotalid sounds not entirely unlike a rattlesnake. The evolutionary implications of this phenomenon are not without interest. Tail vibrating, in fact, is a common nervous habit in many nonvenomous snakes, such as rat snakes and racers.

There are twice as many species of pit vipers in the New World as in the Old—82 against 41. The new-world genera are *Crotalus*, *Sistrurus*, *Agkistrodon* (occasionally still given as *Ancistrodon*, which was undoubtedly originally correct but which has been largely superseded by a spelling error), *Bothrops*, and *Lachesis*. The last-named genus has only a single species, but one of considerable interest.

The old-world genera are *Agkistrodon* (with the same spelling confusion) and *Trimeresurus*. It will be noted that *Agkistrodon* is the only genus found in both the Old and New Worlds and that *Trimeresurus* is the only genus of pit viper not found in the New World.

CROTALIDAE IN THE NEW WORLD

The Bushmaster   The genus *Lachesis* has only a single species, *L. mutus*, the bushmaster. Its maximum length is sometimes said to be 12 feet although that would be truly an extraordinary example. If the bushmaster does in fact reach that length it would exceed the taipan as previously given and make it the third longest venomous snake in the world. Nine feet would be more like a maximum length. It is easily, however, the largest venomous snake in the Western Hemisphere.

A number of factors have conspired to give the bushmaster an evil reputation. First, it is extremely large and that in itself seems to inspire dread when a venomous snake is involved. (Of course, such fear has some basis because there is a general correlation between the size of a snake and the amount of venom it gives. This is rough, however, and not easy to chart.) A second factor is that the bushmaster is notoriously difficult to maintain in captivity. It usually refuses food and only very few specimens have been maintained for any length of time and then, usually, in study collections away from the public. This "seldom seen" factor has made of the bushmaster a large venomous snake from a distant jungle seldom seen, little known, and deadly dangerous. I was told on one occasion that the bushmaster was too dangerous to keep in captivity and that this was why you never saw them. Few herpetologists and zoo directors would agree. They would like few things better than to be able to display a nine or ten foot bushmaster.

The bushmaster (*Lachesis mutus*) is one of the largest venomous snakes in the world, and one of the most overrated. It is, in fact, a mild mannered snake and the statistics from South America rarely carry reports of bushmaster bites.

In fact there seems to be considerable question as to the bushmaster's deadly nature. The snake is without doubt seriously venomous—that certainly. It has long fangs and a good supply of venom, but there are few bites on record. One reason for the sparse bite statistics is probably the fact that the bushmaster is strictly nocturnal and sticks to the forests. Lest there be confusion, the advantage or disadvantage of venomous snakes being nocturnal should be clarified. The nocturnal habits of the African puff adder present a real danger. However, the nocturnal habits of the bushmaster make this snake less of a menace. The difference lies in other habits. The puff adder comes around human habitations seeking rats and gets stepped on. The bushmaster remains in the forests from southern Nicaragua to the lowlands of Ecuador near the coast and in the Amazon basin of Peru, Bolivia, Brazil, and Paraguay. The habits of one assure encounters, while those of the other preclude them.

In addition to being remote and nocturnal, the bushmaster is apparently shy and slow to take offense. While it is not suggested that this almost mythical snake can be handled with impunity, it certainly is not the dreaded monster of fiction. It is potentially extremely dangerous, especially larger specimens, but it has never figured in public health statistics to the degree some other Crotalidae have.

Schmidt (1957) told the story, a humorous story by the greatest of good luck, about some people who were discovered dragging a large bushmaster along a dusty road on a leash they had fashioned from a shoelace. Periodically they would stop and push the reluctant snake along, for it was not very good about being walked like a dog. They thought they had a small boa constrictor. It would have been interesting to see their reactions when they realized what it was they had in their unguarded grips. This story, which is reinforced by other similar tales, seems to reveal the truly placid disposition of the bushmaster.

Rosenfeld (1971) presents data on two series of snakebites in South America. In one series, from 1902 to 1945 there were 6,601 bite cases. Only 16 of these were by the bushmaster (less than a 0.25 percent) and of the 16 only one person died—6.2 percent. In a second series from a hospital in Brazil, for the 12 years 1954-65, there were 1,718 cases of snakebite with not one of them by the bushmaster. Minton and Minton (1969) adopt a word from the German zoologists—*Kulturfliegern*, literally translated meaning "culture fleers"—to describe this snake. Bushmasters are undoubtedly *Kulturfliegern* and that is why they are so seldom involved in accidents with human beings. As for the tales of mayhem in Brazilian jungles, they are tales. This largest of all pit vipers and largest venomous snake of the New World will always remain potentially dangerous and a darling of the fiction writers.

Bothrops Pit Vipers   The genus *Bothrops* is of considerably more concern than genus *Lachesis*. It is widespread in Mexico, Central

America, and South America, in both tropical and temperate zones. The *Bothrops* pit vipers vary widely in their habits, their size, and their toxicity. None are found north of Mexico and none, of course, outside this hemisphere.

The most widely distributed species is *Bothrops atrox*, the barba amarilla or yellow-beard. In many parts of its range (perhaps in most) it accounts for 80 to 85 percent of the snakebites reported. It is found in Bolivia, Brazil, Colombia, Ecuador, the Guianas, Peru, and Venezuela.* *B. atrox* apparently comes in a variety of sizes and color phases. Lengths for average adults are given as from three feet to about five. Maximum reported is apparently eight feet.

Common or vernacular names are, of course, a nightmare of confusion. *B. atrox* has a great variety of names the most popular of which, besides yellow-beard, are terciopelo, or velvet skin, and fer-de-lance. The name fer-de-lance, in turn, is a volume full of confusion in itself. Any number of snakes in Central and South America (and technically in North America since Mexico, too, has species) are called fer-de-lance (but not by the local people). The actual fer-de-lance, as we shall see, comes from the single West Indian island of Martinique, and nowhere else.

Reports of the toxicity of the yellow-beard are absurdly divergent, literally ranging from 1 percent to 100 percent for mortality. In fact, although this snake is very toxic, has long fangs, and probably kills more people in tropical America than any other species of snake, the mortality rate is low, perhaps as low as 2 or 3 percent. It bites such an enormous number of people because it is so common and so widespread that even the low percentage that die constitute the largest harvest of any snake in the hemisphere, or at least in the tropical zones of the New World. The venom is powerfully hemotoxic and gross tissue damage to the limb involved is common. Amputation or permanent mutilation are not uncommon results of an encounter.

*There is some taxonomic confusion. Depending on the authority used, barba amarilla could be said to range north to Mexico.

The cotiarinha (*Bothrops itapetiningae*) is a relatively obscure arboreal snake from Brazil. Although distinctly a venomous snake it probably doesn't figure very highly in human mortality statisitcs.

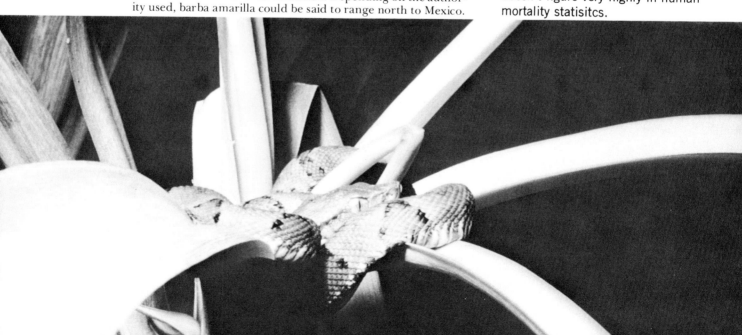

The one and true fer-de-lance is *B. lanceolatus* and is restricted to Martinique. It grows to be seven feet long and is an active, nervous animal. One hears absurd statements of 100 percent mortality for this snake, too, but they are easily recognized as gross exaggerations. The fer-de-lance administers a bite that is both hemotoxic and neurotoxic and is without doubt dangerous, largely because it is likely to attempt to strike when encountered. It pursues rats into cane fields and other agricultural enterprises and is encountered by barefoot workers. The incidence of bites can be high. No clear data is available to give us a mortality rate. It is clear, however, that it is better not to be bitten by the true fer-de-lance of Martinique. Fortunately, this snake now has a much more restricted range than once was the case.

Another widespread member of the genus is *B. schlegeli*, the horned palm viper or, as it is often known, the eyelash viper. It is a very common arboreal snake found in British Honduras, Colombia, Costa Rica, Ecuador, El Salvador, Guatemala, Honduras, Mexico, Nicaragua, Panama, and Venezuela. Paralysis has been reported among the systemic reactions to serious envenomation by this snake. Most victims survive, however.

There are a number of so-called palm vipers, but none are more common than *B. schlegeli*. The rare green or yellow-lined palm viper, *B. lateralis*, is found in Costa Rica and Panama, while the black-spotted palm viper *B. nigroviridis*, another rare snake, lives in Panama, Mexico, Honduras, Costa Rica, and Guatemala. There are others as well. The name is misleading because these snakes, although certainly arboreal, are not restricted to palm trees. They can be found in almost any tree or substantial bush. Their bites can be particularly bad because they are often delivered around the neck or face. Coffee pickers are frequent victims of the arboreal *Bothrops* species.

The fer-de-lance (*Bothrops lanceolatus*) is found only on the West Indian island of Martinique. The many Central and South American species called fer-de-lance are misnamed. The true fer-de-lance is a large, nervous snake and has caused a number of human fatalities.

The jararacussu (*Bothrops jararacussu*), a snake from Brazil, Bolivia, Paraguay, and Argentina. Specimens have been reported to grow to five and a half feet. It is an aquatic snake and although not common, regularly causes serious accidents. An early symptom of a bite is blindness.

The jararaca (*Bothrops jararaca*) of the grasslands of Brazil, Paraguay, and Argentina. It is exceedingly common in some areas and is a major cause of snakebite death where it is found.

The urutu (*Bothrops alternatus*), a species from the waterways of Brazil, Uruguay, Paraguay, and Argentina. It is quick to bite and although local damage can be extensive death among humans is rare.

The barba amarilla or yellow-beard (*Bothrops atrox*) ranges from Mexico to deep into South America. It is probably responsible for more human deaths in tropical America than any other species—and that would mean in the entire Western Hemisphere.

Bothrops neuwiedi, the jararaca pintada, is a three-foot snake with a moderately wide range—northern Argentina, eastern Boliva, southern Brazil, Paraguay, and Uruguay. It comes in a bewildering assortment of subspecies and has an endless number of local or vernacular names. This snake is known to be a cause of some concern in Argentina and is probably a public health factor in a number of other parts of its range as well.

The Amazonian tree viper, *B. bilineatus*, is one of the more widespread of the prehensile-tailed tree vipers and is much less of a hazard than some adventure writers would have one believe. It grows to a length of three feet and although bites are known, deaths from *B. bilineatus* are very few, perhaps absent altogether.

Before going on to the other genera in the family Crotalidae, we should acknowledge a few other of the many *Bothrops* species. *B. jararaca*, the jararaca, is a very common snake found in Argentina, Brazil, and Paraguay. It grows to a length of six feet, making it one of the larger pit vipers. Some deaths have been reported, but massive tissue damage is more likely to be the outcome of its bite. *B. jararacussu*, the jararacussu, is found in Argentina, Bolivia, Brazil, and Paraguay. It is largely aquatic and grows to a length of five and a half feet. Again, deaths are relatively rare although tissue damage can be extensive. When serious systemic reactions occur, blindness is apparently experienced early in their onset.

The urutu of Argentina, Brazil, Paraguay, and Uruguay and the cotiara of Argentina and Brazil are both also members of the genus *Bothrops*, *B. alternatus* and *B. cotiara*. The urutu is capable of causing severe local tissue damage and occasionally death.

The last snake of the genus we will discuss is one of the most interesting—*Bothrops nummifer*, the jumping viper. A stout two-foot snake, it is found in El Salvador, Guatemala, Honduras, Mexico, Nicaragua, and Panama. It is unduly feared probably because it has the uncommon habit of jumping a distance about equal to its own body length at its intended victim. It does what almost all other snakes are said to do, but do not. Although the venom of this snake is slow-acting and deaths resulting from its bite all but unknown, its habit of jumping at people and its other habit of hanging on and chewing have given the jumping viper a rather bad name. Being jumped at by a venomous snake could put a person off somewhat.

These, then, are some representatives of the very large genus *Bothrops*. It is a highly successful genus and has spread through much of the Western Hemisphere. One can speculate and suppose that it is naturally spreading northward and will in time invade the United States, human pressures on habitat notwithstanding. It would, of course, find competition here; for what

Bothrops is to southern Mexico, Central and South America, the rattlesnakes are to North America. We will discuss them next, after listing the antivenins available for *Bothrops*.

ANTIVENIN FOR BOTHROPS

Source	Product Name	Species Covered
Instituto Nacional de Microbiologica (Argentina)	Bivalenta Anti-Botropico	B. alternatus B. neuweidi
	Polivalente	B. alternatus B. neuweidi
	Polivalente Misiones	B. alternatus B. neuweidi B. jararaca B. jararacussu
Instituto Butantan (Brazil)	Soro Anti-Botropico	Bothrops sp.
	Soro Anti-Ophidico	Bothrops sp.
Instituto Pinheiros (Brazil)	Soro Anti-Botropico	B. alternatus B. atrox B. jararacussu B. jararaca B. cotiara
	Soro Anti-Ophidico	B. alternatus B. atrox B. jararaca B. jararacussu B. cotiara
Instituto Nacional de Salud (Colombia)	Suero Anti-Ofodico	Bothrops sp.
Behringwerke AG (Germany)	Serum Mittle-und Sudamerika	B. atrox B. jararaca
Instituto Nacional de Higiene (Mexico)	Suero Antibotropico	B. atrox
	Suero Antiviperino	B. atrox
Laboratorios MYN (Mexico)	Suero Antibothropico MYN Liofilizado	B. atrox
	Suero Antiviperino MYN Liofilizado	B. atrox
	Suero Antiofidico MYN Polivalente (Liquido)	B. atrox
Wyeth Laboratories (United States)	North and South American Crotalid Antivenin	B. atrox
Laboratorio Behrens (Venezuela)	Suero Antibotropico	B. atrox Bothrops sp.
	Suero Antiofidico	B. atrox Bothrops sp.

Rattlesnakes   The rattlesnakes, Crotalidae of the two genera *Crotalus* and *Sistrurus*, have played an extraordinary role in the history, religion, symbology, and legendry of North America. Their epicenter is the southwestern part of the United States and adjacent Mexico. From there they have spread out, becoming less diversified the farther away they get. Very few forms are found in Canada, largely due to the latitude, of course. In South America we again find much less diversification, competition probably having more to do with limiting the snake than geography. *Bothrops*, as we have seen, is the genus of pit vipers that occupies Central and South America. In general, *Bothrops* is less tolerant of aridity and coolness; *Crotalus* of prolonged high humidity.

To give this survey some numerical value, note the diminution of subspecies as one moves away from the epicenter of rattlesnake development:

The Santa Catalina Island rattlesnake (*Crotalus catalinensis*), the only rattleless rattlesnake. It is apparently typical in every other way. Note the clearly visible heat-sensing pits.

Area		Number of Subspecies
Arizona		17
California	*Epicenter*	10
New Mexico		8
Texas		10
Alabama		6
Florida		3
Oregon		2
South Dakota	*Northward*	1
Massachusetts		1
Alberta, Canada		1
Ontario, Canada		1
Baja California, Mexico*		11
Chihuahua State, Mexico*		10
Panama	*Southward*	0
Venezuela		2
Brazil		1
Chile		0

*Until quite recently Mexico was not "as well worked" as the American Southwest. Work is in progress and new species and subspecies are being realigned and validated.

Rattlesnakes have been celebrated by many Americans in addition to the colonists who featured the snake on their flags. The American Indians and those of Mexico and Central America depicted the rattlesnake in their art and in their traditional legends. To some degree the snake was worshipped or at least accredited with supernatural powers. There were parallels with the values placed on cobras in India to this day. But, even the Europeans, once they got here, were not immune to the influence of the rattlesnake.

There is a passage in the New Testament (Mark 16:17-18) that advises that "... they shall take up serpents; and if they drink any deadly thing, it shall not hurt them. ..." Following this line of thinking, a serpent-handling religious cult started up in the American Southeast around the beginning of this century using rattlesnakes, and their close relatives, the copperheads and cottonmouths (genus *Agkistrodon*). Religious ecstasy proved to be poor protection, however. Since the cult first became known outside of its rural seat, 20 people are known to have died from bites suffered during ceremonies. There were probably a good many more the knowledge of which was suppressed to avoid action by lawmakers. Included in the 20 victims we know of was the believed founder of the movement, George Went Hensley.

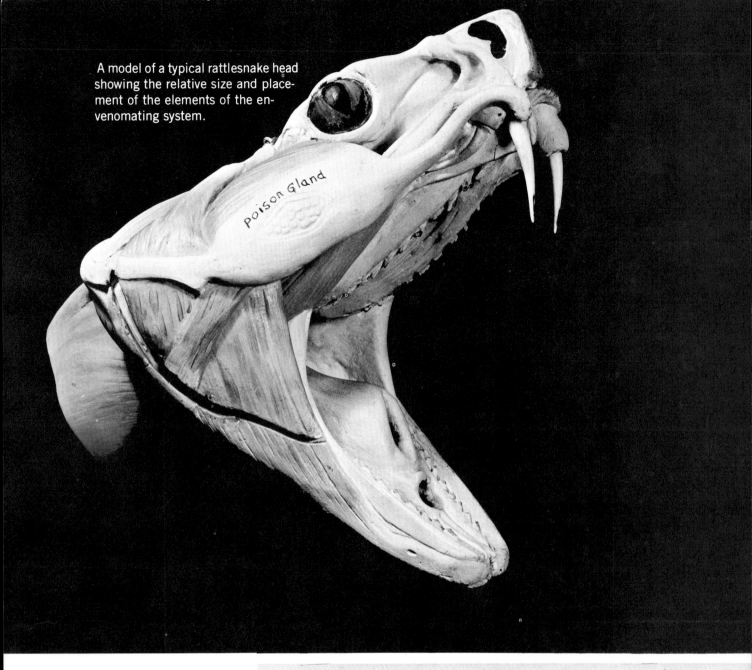

A model of a typical rattlesnake head showing the relative size and placement of the elements of the envenomating system.

poison gland

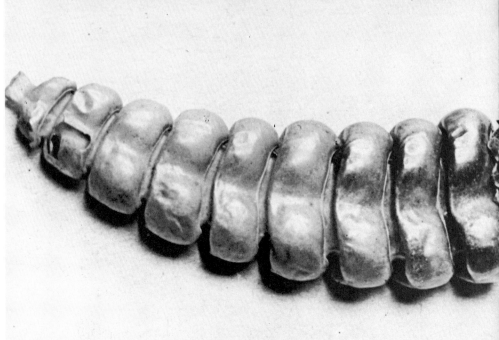

The rattle, a unique warning device not found in any other group of snakes. It is actually a defensive mechanism for the snake itself warning other animals away and obviating the need for useless showdowns resulting from chance encounters.

The way to hold a rattlesnake, if a rattlesnake really must be held. The practice is to be discouraged. A surprising number of accidents result from amateurs handling venomous animals.

The cascabel (*Crotalus durissus*) is the rattlesnake of South America and one of the most dangerous of all rattlesnakes. It ranges from Mexico to deep into South America. Its venom has powerful neurotoxic constituents and an enormous amount of antivenin is required to counteract a bite. Death is not uncommon for a human victim.

The rattlesnakes, their value in religion and other forms of ecstasy aside, are a remarkable group of animals. They have the unique device of the rattle. Its use has long been a subject of debate, and not just a little nonsense. Despite the endless stories to the contrary, the rattle is not for attracting prey, nor for paralyzing prey with fright, but is in fact the rattlesnake's first line of defense. It warns away animals that might harm the snake. Thus, in a land once crawling with wildlife (including an estimated 100 million bison) the rattlesnake was able to avoid being stepped on or taken up as food by a variety of predators. (Not all predators, though, can be warned off.) The fact that a rattlesnake can incapacitate an animal with its bite would not necessarily save the rattlesnake's life. More than one snake has struck out and bitten its foe just before it died.

The envenomating apparatus of the rattlesnakes is essentially a food-getting device, as it is presumed to be with all terrestrial snakes. It is secondarily defensive.

Rattlesnake venom, in most species, is hemotoxic and often powerfully so. The venom of some species, however, also contains a strong neurotoxic constituent (*C. scutulatus* and *C. durissus*) and their bite is often elapid-like in systemic effects. It will be worthwhile here, I think, to mention the origin of fangs and venom. The material injected is essentially a digestive enzyme, which is of great value to an animal whose teeth are too fine to tear or crush food. When a rattlesnake, or any venomous snake, strikes it is injecting a substance that starts the digestive process even as it kills the prey. In a sense, then, neither the snake's teeth nor its venom has been modified in purpose, only in form. It is presumed that the saliva of all snakes contains substances which if injected into a person would be destructive to him. Venom is modified saliva and fangs are modified teeth.

The eastern diamondback rattle-snake (*C. adamanteus*) is considered by some experts to be one of the most dangerous snakes in the world. It is aggressive and can grow to be almost eight feet long. It is from the southeast where its numbers have been reduced in recent years. It is seldom tolerated when encountered.

They are, singly and in combination, the kind of equipment nonvenomous snakes have, only more so. They are extraordinary evolutionary refinements.

Fairley (1929) postulated the following principal categories in the makeup of snake venoms:

> (1) Neurotoxins acting on the bulbar and spinal ganglion cells.
>
> (2) Hemorrhagins destroying the endothelial cells lining the blood vessels.
>
> (3) Thrombase producing intravascular thrombosis.
>
> (4) Hemolysins destroying red blood corpuscles.
>
> (5) Cytolysins acting on blood corpuscles, leucocytes, and tissue cells.
>
> (6) Antifibrins or anticoagulins retarding the coagulation of the blood.
>
> (7) Antibactericidal substances.
>
> (8) Ferments and kinases for the purpose of preparing the prey for pancreatic digestion.

Klauber (1956) suggests that rattlesnake venom is particularly rich in elements 2, 3, and 5 (elapids in 1, 4, and 6). There are species that are strong in the elapid-like constituents, but rattlesnake venom can be fairly characterized as essentially hemotoxic.

Before going on to compare rattlesnakes or even to discuss them, since they are so diverse and far spread, we should introduce the various species. These are identified in the following list:

The large, aggressive, and dangerous western diamondback rattlesnake (*C. atrox*). The species is found from Arkansas to California, from Missouri to well into Mexico. Certain places are densely populated with these snakes.

GENUS *CROTALUS*

Species	Number of Subspecies	Maximum Length Believed Reliable	Range
C. adamanteus Eastern diamond rattlesnake		7'11"	Ala., Fla., Ga., Miss., N.C., S.C.
C. atrox Western diamond r/s		6'11"	Baja, Ariz., Ark., Calif., Kans., Nev., Tex., N.M., Okla.
C. basiliscus Mexican west-coast r/s	2	6'8½"	6 Mexican states
C. catalinensis Santa Catalina Island r/s		2'3"	Santa Catalina Island
C. cerastes Mojave Desert sidewinder r/s	3	2'6"	Ariz., Calif., Nev., Utah, Sonora, Baja
C. durissus Central and South American r/s	6-11*	5'6"	Brit. Hond., Guat., El Salv., Hond., Nic., Costa Rica, Pan., Colom., Venez., Brit. Gui., Surinam, Fr. Gui., Braz., Peru, Para., Urug., Arg., 9 Mexican states
C. enyo Lower California r/s	3	2'11"	Baja, 6 offshore islands
C. exsul Cedros Island diamond r/s		3'	Cedros Island
C. horridus timber r/s	2	6'2"	Ark., Ia., Kans., La., Minn., Mo., Neb., Okla., Tex., Ala., Conn., Fla., Ga., Ill., Ind., Ky., Md., Mass., Miss., N.H., N.J., N.Y., N.C., Ohio, Pa., R.I., S.C., Tenn., Vt., Va., W. Va., Wis., Ont.
C. intermedius Totalcan small-headed r/s	2	1'10"	4 Mexican states
*C. lannomi***			Western Mexico
C. lepidus mottled rock r/s	3	2'8½"	N.M., Tex., Ariz., 8 Mexican states

*Being revised.
**Recently established.

Species	Number of Subspecies	Maximum Length Believed Reliable	Range
C. mitchelli San Lucan speckled r/s	4	4'3"	Calif., Nev., Ariz., Baja, 6 offshore islands
C. molossus Northern black-tailed r/s	3	4'1½"	Ariz., Calif., Nev., N.M., Tex., 18 Mexican states
C. polystictus Mexican lance-headed r/s		3'	13 Mexican states
C. pricei Arizona twin-spotted r/s	2	1'10½"	Ariz., 5 Mexican states
C. pusillus Tancitaran dusky r/s		2'2½"	2 Mexican states
C. ruber red diamond r/s	2	5'4"	Calif., Baja, 5 off-shore islands
C. scutulatus Mojave r/s	2	4'	Ariz., Calif., Nev., N.M., Tex., Utah, 9 Mexican states
C. stejnegeri long-tailed r/s		2'	2 Mexican states
C. tigris tiger r/s		2'10½"	Ariz., Sonora, Mex.
C. tortugensis Tortuga Island diamond r/s		3'5½"	Tortuga Island
C. transversus cross-banded mountain r/s		1'6"	3 Mexican states
C. triseriatus Central-plateau dusky r/s	2	2'2½"	9 Mexican states
C. vegrandis*		3'(?)	Venezuela
C. viridis prairie r/s	9	5'4"	Ariz., Col., Calif., Ida., Ia., Kans., Mont., Neb., N.M., N.D., Okla., Ore., S.D., Tex., Utah, Wash., Wyo., Alta., B.C., Sask., Baja, Los Coronados I., 3 Mexican states
C. willardi Arizona ridged-nosed r/s	3	2'1"	Ariz., N.M., Sonora and Chihuahua, Mex.

Species	Number of Subspecies	Maximum Length Believed Reliable	Range
GENUS *SISTRURUS*			
S. catenatus Massasauga	2	3'1½"	Tamaulipas and Coahuila, Mex., Ont., Ariz., Col., Ia., Kans., Minn., Mo., Neb., N.M., Okla., Tex., Ill., Ind., Mich., N.Y., Ohio, Pa., Wis.
S. miliarius Southeastern pigmy rattlesnake	3	2'6½"	Ala., Fla., Ga., La., Miss., N.C., S.C., Tex., Tenn., Ark., Okla., Ky.
S. ravus Mexican pigmy r/s		2'3½"	7 Mexican states

The widespread prairie rattlesnake (*C. v. viridis*) causes more human and livestock injuries each year than any other species of rattlesnake. In some areas there are very dense populations of these animals. They usually do not cause human mortality.

The Carolina pygmy rattlesnake (*Sistrurus miliarius miliarius*) is a small species from the American southeast—South and North Carolina, Georgia, and Alabama. Human deaths are virtually unknown but the bite can be very painful. The species is reportedly quick to bite.

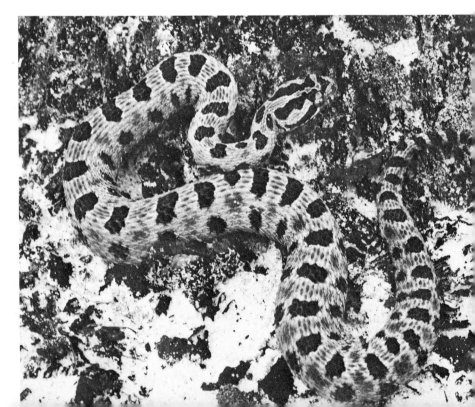

The word *rattlesnake* is used generally as if it were a definitive statement that told a great deal about an animal. In one sense it does, of course. It tells us a species belongs to one of two genera, that it is new-world, that it is venomous, that it has heat-sensory pits, and that it has a horny warning device at the end of its tail. There the generalities start to thin out except for technical details of morphology. Take the subject of range, for example. Some rattlesnake species like the Cedros Island diamond rattler (*C. exsul*) are severely restricted in range while others like the Central and South American rattler (*C. durissus*), the timber rattler (*C. horridus*), and the prairie rattler (*C. viridis*) include tens or even hundreds of thousands of square miles in their range.

When it comes to size we again find enormous differences. There are giants among the rattlesnakes. The Eastern diamond rattler (*C. adamanteus*) at least approaches eight feet in exceptional specimens, while *C. atrox*, its Western counterpart, approaches seven feet exceptionally. The Mexican west-coast rattler, *C. basiliscus*, exceeds six feet eight inches. Compare these monsters with the Totalcan small-headed rattler (*C. intermedius*), the Arizona twin-spotted rattler (*C. pricei*), and the cross-banded mountain rattler (*C. transversus*), none of which reach two feet in length.

The timber rattlesnake (*Crotalus horridus horridus*) is an active and at times rather cranky snake with a range over many parts of the northeastern and north central United States. They can be found within the city limits of even large municipalities.

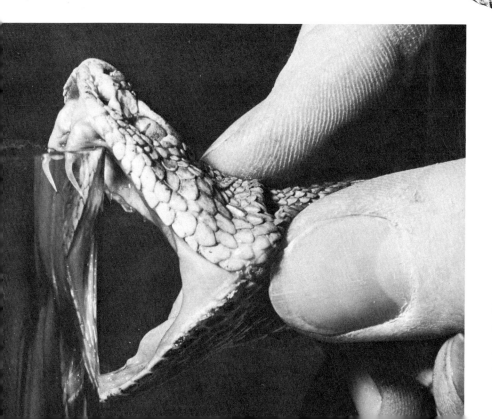

Milking the timber rattlesnake (*C. h. horridus*)—the impressive fangs of this species can be clearly seen. In an actual bite the jaws would be spread even farther apart than this.

The Mexican west coast rattlesnake (*C. b. basiliscus*) is believed to be a very seriously venomous species although little work has been done with its venom. It has been known to grow to a length of six feet.

Rattlesnakes range from the lowlands to the mountains, from semi-aquatic environs to hot sand desert. It is precisely because the two genera have so diversified, so spread out, and so adjusted things like size and habit to take advantage of all opportunities that they are so successful, and the representatives of the many species have so often come into conflict with man.

Rattlesnakes often are found in surprising population densities. In many areas this results, in part, from the snakes' habit of hibernating, and denning while they do so. Absolutely enormous numbers of snakes may use the same cave or hole as a hibernation den and the period immediately before the snakes go into the den (and just after they come out and before they have had a chance to disperse) will see the rattlesnake population of the area very high. In other areas, where the rattlesnakes are active all year round, there still may be a high population density simply because the rattler, a live-bearer, is prolific.

Some indication of this density can be seen around Sweetwater, Texas. The Junior Chamber of Commerce has made capital of the rattlesnakes in that area by staging an annual "Rattlesnake Round-Up." As an event it does attract attention and therefore commerce to the area (Nolan County; Sweetwater has a population of about 14,000). It is instructive to review the catches around Sweetwater since the publicity stunt was instituted in 1959:

SWEETWATER RATTLESNAKE CATCH

Year	Number of Snakes	Total Weight
1959	3,000	3,128 lbs.
1960	6,881	8,989
1961	4,044	4,584
1962	2,392	2,486
1963	5,000	4,500
1964	3,900	3,762
1965	1,900	2,340
1966	4,021	3,400
1967	4,300	4,000
1968	3,400	
1969	3,500	2,474
1970	9,017	8,886
1971	4,246	3,700

There were ten bite cases in the course of these "round-ups," a remarkably low number when one considers that 55,601 venomous snakes (weighing over 26 tons) were handled not just once, but perhaps several times each. It would seem to speak well for the reflexes of Texans, or badly for the reflexes of rattlesnakes.

The canebrake rattlesnake (*C. hor-ridus atricaudatus*) is a southern sub-species of the timber rattlesnake. It ranges west to Texas and can be short-tempered and dangerous.

The midget faded rattlesnake (*C. v. decolor*) is one of the many subspecies of the prairie rattlesnake. It is found in parts of Wyoming, Utah, and Colorado. Human deaths have not been reported.

The Mojave rattlesnake (*C. s. scutulatus*) is, despite its common name, rather widespread—California, Nevada, Arizona, New Mexico, Utah, Texas, and Mexico. It is believed to be one of the most dangerous of all rat-tlesnakes with powerful neurotoxic elements in its venom.

The Cedros Island rattlesnake (*C. exsul*) is found only on Cedros (or Cerros) Island off the Pacific coast of Baja California. The alert posture of this specimen is typical of all rattlesnakes. They seem ever-ready to take offense.

The Arizona ridge-nosed rattlesnake (*C. w. willardi*) is found in the mountains of southeastern Arizona and in northern Mexico. It is an alert and often cranky animal and one not to be taken lightly.

The Arizona twin-spotted rattlesnake (*C. p. pricei*) is a highland species from southern Arizona ranging into Mexico. There are 17 forms of rattlesnake in Arizona making it clearly the rattlesnake capital of the world.

The northern Pacific rattlesnake (*C. v. oreganus*) is one of the more handsomely marked subspecies of the prairie rattlesnake. This form is found up into British Columbia and far into California. Like all of the subspecies of the prairie rattlesnake it rarely causes human deaths. It is, nonetheless, seriously venomous.

The northern black-tailed rattlesnake (*C. m. molossus*) is found in Arizona, Texas, New Mexico, and Mexico. It is a handsome but dangerously alert animal that is usually active at night, like most snakes of the region.

The eastern massasauga (*S. c. catenatus*) grows to be just a little over two feet long. It is retiring but will bite if angered. The bite can cause human death although that would be an unlikely result.

The Mojave Desert sidewinder (*C. cerastes cerastes*) is a desert rattlesnake of the American Southwest. Active only at night it uses a peculiar sideways coiling movement to progress across the sand. It can be encountered in caves and even old buildings where it retreats from the heat of the day.

The red diamond (or diamondback) rattlesnake (*C. r. ruber*) from California and Baja California. Despite the menacing appearance in this close-up, this species is less aggressive and has milder venom than either of the other two diamond rattlesnakes. It does grow to about five feet, however, and it would be foolish indeed to depend on such a snake's mild disposition.

Bites by rattlesnakes are often characterized by severe local tissue damage and amputations are not uncommonly required. Note this case report:

Twenty-two-year-old male bitten in web of left thumb (one fang only). Five foot rattlesnake identified positively as culprit. On admission to hospital 30 minutes later he was in shock, was incontinent, vomiting blood, and complained of severe visual deficiency. After 24 hours there was marked edema, and ecchymosis (purple discoloration from blood seeping out into tissues) extending into the chest wall. He experienced difficulty in urinating on the second day. After 72 hours the hand appeared gangrenous. On the fifteenth day his hand had to be amputated—mid-forearm [Adapted from McCullough and Gennaro, 1963].

This case is not an atypical example. Many of the venoms of North American snakes exhibit such a strong necrotizing effect. Sherman Minton has suggested that there may be an inverse relationship between the lethal quality of a venom and its ability to cause necroses. This is a relationship that will be profitable to pursue, assuming that it does in fact exist.

Wood, Hoback, and Green (1955) proposed a classification for snakebite intensity:

Grade 0: NO VENENATION. Fang or tooth marks, minimal pain, less than 1 inch of surrounding edema and eythema. No systemic involvement.

Grade I: MINIMAL VENENATION. Fang or tooth marks, severe pain, 1-5 inches of surrounding edema and eythema in first 12 hours after bite. No systemic involvement usually present.

Grade II: MODERATE VENENATION. Fang or tooth marks, severe pain, 6-12 inches of surrounding edema and eythema in first 12 hours after bite. Systemic involvement may be present, nausea, vomiting, giddiness, shock, or neurotoxic symptoms.

Grade III: SEVERE VENENATION. Fang or tooth marks, severe pain, more than 12 inches of surrounding edema and eythema in first 12 hours after bite. Systemic involvement usually present as in Grade II.

Many investigators suggest that between 20 and 25 percent of all rattlesnake bites fall under Grade 0, involving little or no envenomation or *venenation*. The grade under which the remaining 75-80 percent of the bites fall depends on a number of factors. These can include species, size, health, and recent activities of the snake, as well as its disposition, that is, how seriously it has been annoyed or provoked. Human factors include age, weight, health, susceptibility to foreign proteins (toxins), general allergy profile, recent medical history, and even psychological factors, it may be assumed. Other factors include location of bite, whether one or two fangs were involved, the actual amount of venom injected, whether the fang broke off or remained intact, whether the snake bit more than once (unusual in rattlesnakes), and, of course, first-aid course initiated and medical treatment given. The possibility of later infections and of systemic reaction to serum therapy can all play a role in the chances of survival.

It should be obvious from both the diversity of the rattlesnakes themselves and from the number of factors governing snakebite severity that there is no way of characterizing a rattlesnake bite except in the most general terms. These could be:

A. A rattlesnake bite is likely to be represented by puncture wounds (one or two) that are relatively deep for the size of the snake involved. (Solenoglyph fang type.)

B. The venom injected is more likely to be predominantly hemotoxic and peripherally neurotoxic than vice versa.

C. In cases of severe envenomation significant short-term and long-term damage to local tissue may be anticipated.

D. There will be considerable pain and shock connected with the episode.

E. The bite will probably be on an extremity. There is an almost 85 percent likelihood that it will be on the foot or lower leg.

F. A fair percentage of the bites will result from the handling or attempted handling of a snake. These bites will most likely occur on the hand or wrist.

G. Except in rare cases where long fangs penetrate deeply enough to engage a blood vessel, sudden or quick death may not be expected.

H. Death may be expected in only a very small percentage of the cases. Most serious long-range effects will involve amputation or permanent loss of use of a member.

These same observations will be largely true of all of the Crotalidae, with a notable exception being the cascabel (*C. durissus terrificus*), the South American rattlesnake. Elapid-like neurotoxic reactions will be noted in the clinically-seen case, and mortality will be higher.

The Aruba Island rattlesnake (*C. unicolor*) has one of the most restricted ranges of all rattlesnakes—just that single island off the coast of Venezuela. It is a small snake, two to three feet on the average, and little is known of its venom's strength.

Special note should be made in passing of the Mojave rattlesnake. This four-foot snake is apparently much more seriously venomous than any of its congeners. In 1965, Dr. Fred Shannon was bitten on a finger of his left hand by a Mojave rattlesnake he was attempting to collect. He was only forty-four years old at the time and he died despite the efforts of Dr. Findlay Russell with whom he had coauthored a book on snakebite for the armed forces. Dr. Shannon was considered one of the world's foremost authorities on snakebite.

AGKISTRODON

With genus *Agkistrodon* we will complete our review of the venomous snakes of the Western Hemisphere and bridge ourselves back to the Old World. This genus is found in North and Central America and in Asia as far west as European U.S.S.R.

The pit vipers of this genus have been called rattlesnakes without rattles. There is more fact than fiction to that statement although the *Agkistrodon* are nowhere near as diverse as the rattlesnakes, and smaller on the average than the larger rattlers. In North America north of Mexico the *Agkistrodon* are limited to:

> *A. piscivorus* (cottonmouth water moccasin): to six feet exceptionally. Virginia to Florida and west to central Texas. Wetlands only.

> *A. contortrix* (American copperhead): to four feet exceptionally. New England to Kansas and Texas, but not peninsular Florida.

The cottonmouth water moccasin (*Agkistrodon piscivorus*), a common species in our southern waterways and a cranky animal that can be quick to bite. There can be severe local tissue damage as a result of a bite and deaths have been reported.

Both of these snakes are responsible for a fair percentage of the bites experienced in the United States each year. There is some evidence that the fortunately mildly venomous copperhead bites more people than any other species on this continent. It has been estimated (Parrish, 1967) that the annual number of copperhead bites may be as high as 2,920. The mortality rate is put at 0.01 percent, one of the lowest (perhaps *the* lowest) for any snake whose bite is a major factor in the public health statistics for a region. Even if the copperhead rates second in the number of bites per year, following collectively the rattlesnakes, it is a leading cause of snakebite while being a nil factor in death statistics.

The cottonmouth probably accounts for about 10 percent of the snakebites in America. When one compares the limited range of this snake to the territory covered by the diverse rat-

The copperhead (*Agkistrodon contortrix*) in normal and unmarked "sport" forms. This is a species that bites many people each year but almost never causes a human fatality.

tlesnakes that figure looms somewhat larger. Whereas long-term effects from copperhead bites are rarely reported, cottonmouths characteristically cause the same kind of tissue damage reported for rattlesnakes, and there are strong similarities between the two snakes' venoms. Witness this unfortunate case report (adapted from Parrish, 1967):

> A twenty-one-year-old boy was bitten on his left lower leg by a cottonmouth while fishing. He was examined by a physician two hours after the bite accident. There were two fang punctures on the lower one-third of the left leg with edema and erythema of the foot and lower leg. The patient complained of severe pain, weakness, and excessive thirst. Even after treatment—which included suction, plasma, whole blood transfusions, and antivenin—the patient remained in serious condition with edema involving the entire left leg and extending into the left lateral flank of the trunk. There was extensive necrosis of the left lower with exposure of tendons. Also, the leg wound appeared to be infected. Six days after the bite the patient developed a headache, stiff neck, and restlessness. Soon afterwards he had trismus, difficulty in swallowing, and convulsions. *Clostridium tetani* were cultured from the wound. The patient expired 2 days after the onset of symptoms of tetanus.

The course this case might have taken without the additional complication of tetanus can not be stated with certainty. The three "A's" are prescribed in snakebite cases—antivenin, antitetanus, and antibiotics. Mortality rate for bites by the cottonmouth water moccasin is put at 0.13 percent by Parrish and Donnell (1967).

The cantil (*Agkistrodon bilineatus*) of Mexico. It is an aquatic snake, rather to Mexico what the cottonmouth is to the United States. Its bite is known to be serious but is probably fatal to humans only in exceptional cases.

The only representative of the genus *Agkistrodon* in the West-ern Hemisphere south of the United States is limited to Mexico and swampy coastal areas of Central America south to Guatemala and the east coast of Nicaragua. It is the cantil (*A. bilineatus*). Exceptionally, this snake may reach four feet, but that is an outsized specimen. In habit it resembles the cottonmouth in a number of ways and it is presumed dangerous. Reportedly, local reactions are much more severe than systemic ones and they appear to be essentially hemotoxic as might be expected.

ANTIVENIN SPECIFIC TO CROTALUS, SISTRURUS, AND NEW-WORLD AGKISTRODON

The genera *Crotalus, Sistrurus,* and *Agkistrodon* are well covered by antivenins:

Source	Product	Species Covered
Instituto Nacional de Microbiologia (Argentina)	Monovalente Anti-Crotalico	*C. durissus terrificus*
	Polivalente	*C. durissus terrificus*
	Polivalente Misiones	*C. durissus terrificus*
Instituto Butantan (Brazil)	Soro Anti-Crotalico	*Crotalus* sp.
		Crotalus sp.
	Soro Anti-Ophodico	*Crotalus* sp.
Instituto Nacional de Salud (Colombia)	Suero Anti-Ofidico	*Crotalus durissus* ssp.
Behringwerke AG (Germany)	Serum Mittle- and Sudamerika	*C. durissus terrificus*
Instituto Nacional Higiene (Mexico)	Suero Anti-crotalico	*C. durissus* ssp.
		C. basiliscus
	Suero Anti-viperino	*C. durissus* ssp.
		C. basiliscus
Laboratorios MYN (Mexico)	Suero Anti-crotalico MYN Liofilizado	Mexican *Crotalus*
Laboratorios MYN (Mexico)	Suero Anti-viperino MYN Liofilizado	Mexican *Crotalus*
	Suero Antiofidico MYN Polivalente (Liquido)	Mexican *Crotalus*
Wyeth Laboratories	North and South American Crotalid Antivenin	Good for all North American snakes of these species, plus *C. durissus terrificus* and at least one *Bothrops* sp.
Laboratorio Behrens (Venezuela)	Suero Anti-crotalico	*C. durissus* ssp.
	Suero Antiofidico Polivalente	*C. durissus* ssp.

The vast Pacific Ocean is crossed by the genus *Agkistrodon* (figuratively speaking, of course, for although often aquatic *Agkistrodon* is never marine) where it is joined by one other genus of pit viper, *Trimerresurus*. It is often stated that *Trimeresurus* is to the Old World what *Bothrops* is to the new. The rattlesnakes have no opposite numbers.

Trimeresurus is much the more widely diversified genus of the pit vipers in Asia. In Indonesia and the Philippines there is one species of *Agkistrodon* to ten of *Trimeresurus*. The other comparisons are, region for region:

Region	Trimeresurus Species	Agkistrodon Species
Southeast Asia	17	4
Eastern Asia	11	5
Central Asia	13	2
Western Asia	0	1

Altogether, there are about 40 species of pit vipers in Asia. They range from the quite small to the impressively large, from under one foot to well over seven.

As it is with all snake venoms it is an oversimplification to state that one is hemotoxic *or* neurotoxic, but we can say the Asian pit vipers, like virtually all of the new-world pit vipers, have predominantly hemotoxic venoms. The pit vipers in some parts of Asia (Hong Kong and Malaya are two examples) cause the bulk of the snakebite cases seen in hospitals, but contribute little or nothing to the mortality figures. It is unlikely that any pit viper in Asia has a mortality rate higher than 1 or 2 percent.

In *Poisonous Snakes of the World* (Department of Navy; Moore, Minton, Dowling, and Russell) there is a suggested breakdown of types for the Asian lance-head vipers (all Asian pit vipers are essentially lance-headed and are often referred to that way):

Group I: Large, long-bodied, long-tailed terrestrial

Group II: Small, short-tailed, short-bodied terrestrial

Group III: Small, moderately long-bodied arboreal with prehensile tails

All of the snakes we will be discussing do fall into these categories. The confusion over vernacular names is extreme, with as many as six or eight species being called by the same English common name.

What might be said for one Asian pit viper will be likely to hold true for several or more. They are closely allied and resemble each other in toxicity, habits, and disposition. These, though, are the more significant or interesting species:

Agkistrodon acutus, the sharp-nosed pit viper, ranges through Taiwan, East China, and parts of Indochina (including Viet-

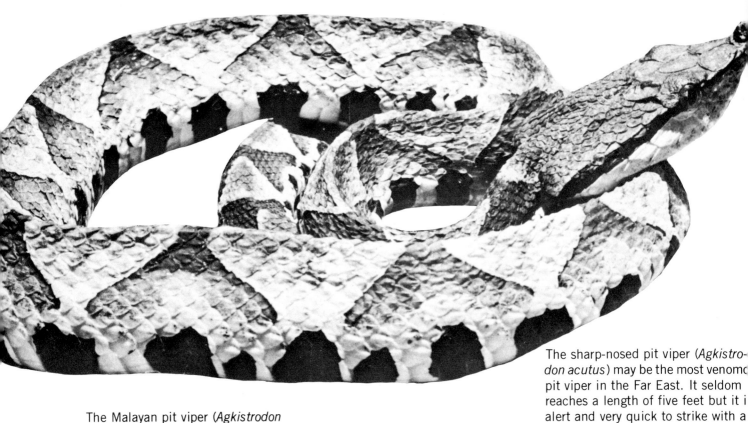

The sharp-nosed pit viper (*Agkistro-don acutus*) may be the most venomc pit viper in the Far East. It seldom reaches a length of five feet but it i alert and very quick to strike with a highly toxic venom.

The Malayan pit viper (*Agkistrodon rhodostoma*), an ill-tempered species from Thailand, Malaysia, Cambodia, Laos, Viet Nam, Java, and Sumatra. Bites are common; human mortality is probably around two percent.

nam). It grows to an impressive length for an Asian lance-head, about five feet in exceptional specimens, and is quite possibly the most seriously venomous pit viper in the Far East. Equipped with long fangs and an irritable disposition, *A. acutus* is certainly a dangerous snake and is quick to strike. Two of its many vernacular names translate as the "5-pace snake" and the "100-pace snake" putting it on a par with the "seven stepper," the common krait in India. A victim of this viper is not inevitably going to die after taking either five or 100 steps, although he will be sick and in pain shortly after the longer of the two walks. (There were six deaths in 37 cases on Taiwan, a mortality rate of 16 percent.) *A. acutus* provides an unpleasant experience for agricultural workers who come too close, and it does kill.

A. halys is known by scores of names over its enormous range. It has seven subspecies and they range from the Pacific Ocean all the way to Europe. In Japan it is called the mamushi, in European Russia, Pallas' viper. It seldom exceeds three feet in length and the mortality rate is extremely low, not more than one victim in a thousand is likely to succumb. Bites, however, are common. In Japan alone between 2,000 and 3,000 people a year are bitten. Many of them are handling the snake at the time of the accident, for it is used as food and in a fierce concoction known as "mamushi whiskey." I have seen drugstores with one window devoted to bottled and packaged medicines and toiletries, many bearing American and British labels, and a second window filled with squirming mamushi waiting to be ground up into one remedy or another. I have watched men and women handle these snakes with less care than a child in New England would show for a garter snake. The fact that the number of bites in Japan does not exceed 3,000 per year attests to the extraordinary nature of these animals. They must be as docile as cocker spaniels. Tens of thousands of them are handled each year. Still, any venomous snake that bites 3,000 people a year in any country is a snake to be reckoned with.

A. rhodostoma, the Malayan pit viper, is an evil tempered snake with a large range, including Thailand, Malaysia, Cambodia, Laos, Vietnam, Java, and Sumatra. It grows to a length of nearly three and a half feet and is said to have a mortality rate among its human victims of 1 to 2 percent. It is a sedentary snake and like many of the Asian lance-heads does not move away when disturbed. It holds its ground (rattlesnakes and cottonmouths often do this) and strikes anything that comes within range. Rubber tappers are often victimized. In Malaya alone, between 1955 and 1960, 23 people died from the bite of this snake. Given the low mortality rate it is evident how many people must be bitten.

In one series of snakebite cases (Reid *et al.*, 1963) the medical records for 384 episodes were summarized. Two hundred eighty-one of the victims had encountered *A. rhodostoma*. Of the 250 selected for review fully half received slight or no venomation. A third of the victims experienced moderate reactions and only a sixth could be characterized as severe cases. The local symptoms included swelling, some pain, extravasation, and blisters, and 11 percent suffered necrotic tissue damage at or near the site of the bite. In severe cases with pronounced systemic reactions there was a general hemorrhagic syndrome up to and including cerebral hemorrhage. Shock is present in the more severe cases and the first sign of a systemic reaction is blood in the sputum. It should be noted that this blood comes from the lungs and not the gums. There can be a prolonged defect in blood coagulation.

A major cause of snakebite in a number of parts of Asia (including parts of India, China, Taiwan, and Vietnam) is the white-lipped tree or bamboo viper, *Trimeresurus albolabris*. Also called the green bamboo snake, it ranges from two to three feet in length and is only mildly toxic. It is nocturnal in habit and during the day is very sedentary. Since the snake often lies coiled in a low branch or in a bush and does not give way when people approach, bites are common. Although medical attention is required they are rarely fatal accidents.

The most impressive of the Asian lance-heads is the Okinawan habu, *T. flavoviridis*. It grows to a reported length of seven and a half feet, making it easily the largest of the Asian pit vipers and is close to being the largest of all pit vipers. It is found in the Ryukyu Islands and in some areas is quite common. It is active at night. (Twice I encountered specimens crossing a paved road late at night on Okinawa.) Bites are common among rural people and the snake's hemotoxic venom is believed to have a mortality rate of about 3 percent. Between 6 and 8 percent of the people bitten are said to suffer some permanent disability. Before World War II horror stories about this snake abounded and there was no small amount of concern among military medical authorities as the Okinawan campaign was mounted. Still, only nine American soldiers were bitten and none of these failed to return to active duty. The stories, as might be expected, were gross exaggerations. The habu is a large snake and a dangerous one, but like most of the Crotalidae it delivers a sublethal bite.

In peninsular India there is a 30-inch snake that is the cause of some problems. The Indian green tree viper, *T. gramineus*, bites often, but kills almost never, except in cases involving small children. Locally it is much feared, but the danger is largely exaggerated. Chaudhuri and his associates (1971), in postulating lethal doses of snake venom for man, created these comparisons: Russell's viper 2.3 times as toxic to man, common cobra 6.6 times, banded krait ten times, saw-scaled viper 20 times, and the common krait 100 times as dangerous to man.

There are more than a score of others. The Chinese mountain viper (*T. monticola*) is a widespread snake that grows to be four feet long (it is usually under two feet, however) and is very defensive. It is a sluggish but irritable animal and its bites are common in many areas. *T. wagleri*, Wagler's pit viper, grows to almost three and a half feet and although mildly toxic does bite many people in the Philippines, Thailand, Malaysia, Indonesia, and Borneo. It is sluggish and nocturnal but during the day, if stumbled upon, will strike defensively.

With this partial review we conclude our discussion of the pit vipers of Asia (and the family Crotalidae) except to note the antivenins available to counteract the effects of their venoms. On the snakebite statistical tables the pit vipers of the Old World rate high in incidence, low in consequence. Still, they make many people ill every day of the year.

ANTIVENINS FOR TRIMERESURUS AND AGKISTRODON IN ASIA

Source	Product Name	Species Covered
Institut Pasteur (France)	Serum Antivenimeux Ancistrodon	*A. rhodostoma*
Perusahaan Negara Bio Farma (Indonesia)	ABM Antivenin	*A. rhodostoma*
Institut d'Etat des Serums Et Vaccins Razi (Iran)	Polyvalent Antivenin for Iran	*A. halys*
Institue for Medical Science (Japan)	Mamushi-venom Antivenin	*A. halys*
	Habu-venom Antivenin	*T. flavoviridis*
Laboratory for Chemo-Therapy and Serum Therapy (Japan)	Antivenin Habu	*T. flavoviridis*
Takeda Pharmaceutical Co. (Japan)	Antivenin Mamushi	*A. halys*
Taiwan Serum and Vaccine Laboratory (Taiwan)	Agkistrodon Monovalent	*A. acutus*
	Hemorrhagic Polyvalent	*T. mucrosquamatus* *T. stejnegeri*
Queen Saovabha Memorial Institute (Thailand)	Antivenine Serum "Green Pit Viper"	*T. popeorum*
	Antivenine Serum "Pit Viper"	*A. rhodostoma*
Wyeth Laboratories (United States)	North and South American Crotalid Antivenin	Not specific to any Asian species, but reputedly of some value in many cases where nothing else is available.

It is difficult to say whether the snakebite problem is exaggerated or unjustifiably minimized. For those people not exposed to the venomous snakes, of course, the problem is exotic, remote, and only half comprehended. For many more millions upon millions of people who make their livelihoods close to the soil in tropical and warm temperate areas the problem is neither academic nor exotic. It is an ever-present reality for them. Although these people do not huddle in their shacks at night in terror of snakes, they must at least be aware of the danger snakes can pose.

Vast quantities of antivenin are produced in the world today specific to the more troublesome species. Many of the polyvalents prove useful in treating bites from species whose venoms are similar to those used in the production of the medicine. The list of species actually covered by the monovalent and polyvalent antivenins now in production will probably not be enlarged appreciably in the years ahead. There may be a few additions, but certainly not many.

Venoms are being studied in many places now and as their structures become known their constituent factors may play an increasingly important role in medicine, not just in the treatment of snakebite, but of many conditions that respond to the kind of effects powerful hemotoxins and neurotoxins can have in controlled situations. The treatment of intractable pain with cobra venom is now decades old. Cobra venom in combination with formic and silicic acid has been used on arthritis patients for a long time. This is a field of medical research in which we can anticipate expansion.

It will be noted that nowhere have we discussed the treatment of snakebite. The literature in this field is vast and not always lacking in confusion. A snakebite, a bite by any snake from a genus known to have venomous species, is a medical emergency and professional medical attention is indicated. Broad spectrum antibiotic programs are needed, protection against tetanus, as well as serum therapy. However, these are matters to be evaluated by a medical professional, including an evaluation of the victim's sensitivity to horse serum. A person suffering from a snakebite who happens to be sensitive to horse serum has enough to worry about without an amateur practitioner coming along and plunging ten cubic centimeters of horse serum into his arm. That can be more deadly than snake venom, and a great deal quicker acting.

Snakebite will continue to be a health problem for at least the balance of this century in many parts of the world. However, as primitive habitats are reduced, as agriculture replaces wilderness and industry replaces agriculture, the snakes will go. There are already vast areas where venomous snakes were once common where they are no longer found. Every venomous snake has a price on its head. The snakebite problem will one day be phased out by a burgeoning human population. No one seems to care to predict what we will do with the annual crop of tens of thousands of tons of rats and mice which will be the snakes' bequest to us.

Chapter 15

THE

VENOMOUS
MAMMALS

The duckbilled platypus (*Ornithor-hynchus anatinus*), a primitive, aquatic, egg-laying mammal from Australia. The male has venomous spurs on the inside of the hind legs. It can be a painful intoxication but human deaths have not been recorded.

 *he Platypus* There is almost nothing about the duckbilled platypus (*Ornithorhynchus anatinus*) that is not surprising. It is a mammal yet it lays eggs. It has fur (it was once hunted as a fur-bearing animal, but is now wholly protected) yet it has a bill that is at least superficially duck-like. It has no external ear, virtually no neck, is amphibious, and both its urogenital organs and shoulder structure are reptilian in character. Its feet are webbed and although it feeds its young with milk the female has no nipples. The milk oozes from ducts in the regions where the nipples might normally be.

The platypus is in the subclass Prototheria, in the order Monotremata. It shares this side limb of the mammalian tree with the five echidnas, or spiny-anteaters, the only other living monotrenes. The five echidna species belong to two genera—*Tachyglossus* and *Zaglossus.* What possible relationship the echidna and the platypus have to the other mammals is not known. In fact, very little is known about their history at all. The fossil record is disappointing. All of the monotrenes come from Australia, Tasmania, New Guinea, and nearby islands.

The male platypus is equipped with a true venom and an efficient apparatus for its use. The animal kicks to envenomate. On the inside of the male's rear legs, down near the heels, are spurs. They are horny, sharp, tapered, conical, and slightly curved. Normally, they are carried in against the legs, but they can be erected voluntarily to a position approximately 90 degrees to the leg and are then driven inward in a convulsive little jabbing motion. Characteristically the platypus uses both legs, catching its opponent with both spurs at the same time. There have been a number of instances where people handling captive specimens were kicked and spurred and *not* envenomated. This *may* mean that the discharge of venom is a separate, subsequent, and voluntary act.

The venom spur is also present in *immature* females, but by the time the animal reaches maturity the spur has vanished. A slight indention is left on the legs where the spurs would be. No case of a female having venom has been recorded, a fact which will undoubtedly be significant in determining why this strange capability evolved.

The spur is connected by a single duct to a reservoir slightly further down the leg. From this reservoir a much longer duct runs up the animal's leg to the inside of the thigh where the kidney-shaped *crural,* or venom gland, is located.

One mysterious aspect of the platypus's venom system is that it is apparently seasonal. Some writers have insisted that this gland becomes enlarged during the breeding season, but others have insisted that caution should be exercised in suggesting any such correlation. There is too little data for a position to be taken.

The question must inevitably arise—What is the platypus's venom apparatus for? The venom itself, the glands that produce it, the erectile spur, and the whole mechanism that supports it are all very specific. They are not *slight* adaptations on other organs or substances. One must assume they had a very good reason for coming into being and it could be further argued that they must still serve a good purpose or they would not be functional today. In plain fact we do not know what they are for, primarily. If it is true that the crural glands swell and become more efficient during the breeding season, then the system may be offensive, being used against other males in establishing a territory or claiming a mate. This would explain why only males are so equipped, and may find support in at least one report that suggests a male platypus killed another male using its spurs.

It has also been suggested by some observers that the spurs are used by the male for holding the female during copulation. This could be, too, but it would not explain the venom, of course. A third suggestion is that the spurs and venom are used to immobilize larger prey. Since the prey of the platypus is seldom

A male platypus with spurs visible. These spurs clearly distinguish the male from the female.

bigger than a frog and is probably more usually insect larvae, this suggestion would seem to describe, at most, a secondary use.

Perhaps all of these things are true, though. Perhaps the male platypus uses his venom to fight off other males, perhaps the spurs are used to hold a female during the act of breeding without the venom being discharged (this would support the idea that venom discharge is a voluntary act not necessarily linked to the erecting of the spurs), and perhaps the spurs are also used on prey. But, what was the primary purpose of the platypus's venom system? What set this strangest of all mammals off on its strange and so-far inexplicable evolutionary tangent?

The venom of the platypus, like all venoms, is imperfectly known. It has been identified as a solution of largely albuminous proteins. There is some proteose present. In its effects it has been described as somewhat like a feeble viper venom. One researcher has suggested that seriously potent snake venom is as much as 5,000 times as toxic to man. It is apparently not neurotoxic. In ten cases of human envenomation the symptoms reported were immediate, intense pain, swelling progressing up the arm, numbness around the wound, some swelling of the glands, and a feeling of faintness. All cases have involved the hands, wrists, and arms and there have been no known human fatalities. In every recorded case the animals were being handled after intentional or inadvertent capture.

David Fleay, an Australian who has been studying the platypus for three decades, has been envenomated on three occasions and spurred on a number of occasions when no venom was injected. He has probably been directly involved with the platypus envenomating apparatus more often than any other person and appears to have suffered no long-term effects.

On occasion dogs have been spurred while attacking, investigating, or retrieving platypuses and there have been reports of some of them dying as a result of envenomation. These reports, however, are often old and fractional and not necessarily reliable.

There has been some work done on the physiological and pharmacological properties of platypus venom but nowhere near enough to shed much light on what is still rather a mysteri-

The echidna or spiny anteater (*Tachyglossus aculeatus*) is an egg-laying mammal. Reports of it being venomous are open to serious doubt. More work is needed to determine if it is and if that phenomenon could be seasonal. It is the kick and not the bite of this animal that would be venomous.

ous capability. We do not know why the platypus developed this unique system of defense, offense, or food-getting, we do not know if it is linked to sexual cycles, and we know very little about the chemistry of the venom itself. Our position, in fact, is one of profound ignorance.

THE ECHIDNA

The echidna, too, apparently has spurs that could be part of a venom apparatus. If possible we know less about the echidna and its venom than we do about the platypus. It is a matter of record that the echidna's spur is similar to that of the platypus, but somewhat smaller, as is the crural gland. It is situated farther down the leg, at the back of the knee. Some investigators suggest that the echidna is not venomous which, if true, could make the gland system attached to the spurs rather difficult to explain unless it is vestigial and in the process of being phased out. If that is the case the question arises, "What in the echidna's environment has changed?" It must take a good deal of impetus to cause such a system to evolve. Where has that impetus gone?

VENOMOUS SHREWS

No one knows how many mammals have a genuinely venomous bite, but there are apparently more than was thought even a few years ago. Up until as recently as 1942 references to venomous shrews were generally laughed off or consigned to the dubiously distinctive category of folklore. This since they have been recorded for centuries as an ingredient in magic brews. Then Dr. O. P. Pearson published what could be accepted as conclusive findings in the *Journal of Mammology*. White mice injected with a substance prepared from submaxillary glands of the North

The short-tailed shrew (*Blarina brevicauda*) with a mouse it has killed by administering a venomous bite in the region of the head and neck. The venom of this little mammal is harmless to man.

American short-tailed shrew (*Blarina brevicauda*) died of true envenomation. Since the shrews were so small there was no practical way of "milking" them the way snakes are "milked." An alternative method of obtaining the venom had to be developed. The submaxillary glands were removed and ground into a powder in a salt solution, using fine sand as an abrasive. It was found that the glands could be used from freshly killed specimens and prepared right away or dried and stored after treatment with acetone. If refrigerated after preparing, it was subsequently found that the venomous preparation could be stored for ten years without losing its toxic qualities.

For a long time, it was assumed that the short-tailed shrew alone of all mammals had a venomous bite. In recent years, however, it has been demonstrated that the European water shrew (*Neomys fodiens fodiens*) and the British bicolored water shrew (*Neomys fodiens bicolor*) are also venomous although not as seriously so as the short-tail. There is every reason to believe that the masked shrew (*Sorex cinereus*) is also venomous.

The shrews belong to the family Soricidae in the order Insectivora. One other family within the order has also been found to contain at least one venomous species. The solenodons (family Solenodontidae) are primitive insectivores limited in distribution to Haiti and Cuba. One species, the squirrel-sized Haitian solenodon (*Solenodon paradoxus*), largest of the North American insectivores, is also demonstrably a venomous animal.

The insectivores, generally mouse-like creatures, are found in at least some part of every continent except Australia and Antarctica. (There are few in South America, none in Greenland.) There are 71 genera but so far only three or at the most four have been found to contain truly venomous animals: *Blarina, Neomys, Solenodon,* and probably *Sorex.*

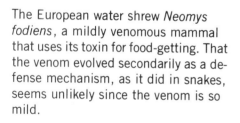

The European water shrew *Neomys fodiens*, a mildly venomous mammal that uses its toxin for food-getting. That the venom evolved secondarily as a defense mechanism, as it did in snakes, seems unlikely since the venom is so mild.

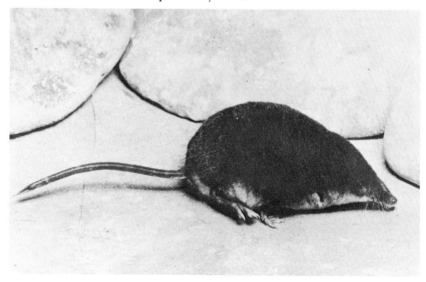

The insectivores are small, inconspicuous, and highly secretive animals and they often escape attention. Since the number of species is large (approximately 300) it is safe to assume that only a very small percentage has been really studied for possible venomous potential. It is difficult to believe that more will not be found to have this strange power as further research is done.

The envenomating apparatus of the Haitian solenodon is relatively simple. The submaxillary glands, well back in the jaw, have two sections, both oval shaped. They are enlarged and may weigh between three and four grams. There are secretory ducts containing large granular-filled cells with small nuclei; the same phenomenon is seen in the European water shrew and the short-tailed shrew. These cells apparently secrete the venom. There are no provisions for venom storage in any appreciable quantity. Ducts lead forward from the submaxillaries and open out near the base of large second incisors which have deep channels. The solenodon bites its victim and venom-bearing saliva is carried along the channels of the second incisors into the wound. The movement of the venom up the channels after discharge from the ducts is undoubtedly caused by capillary action.

In the *Neomys* shrews the system is similar except that the *first* lower incisors are involved and they are not channeled as in the solenodons. They are concave on their inner surface and the venom-bearing saliva collects there and is conveyed to the wound. In *Blarina* the median pair of lower incisors project forward forming a groove between them and it is apparently along this groove that the venom flows.

Although the method of envenomating prey is approximately the same in all three genera, there is a marked difference in the strength of their venoms. *Blarina* has the most powerful. It is suggested that the venom of *Neomys* is about one-third as toxic and that of *Solenodon* about one-twentieth the strength.

Preparations made from the submaxillary glands of all three groups have been tested on mice and other animals. Injections just under the skin cause marked local reactions while injections made intraperitoneally, in the body cavity, and intravenously kill the animals in anywhere from three to 19 minutes, depending on dosage and the weight of the animal being injected.

Although a short-tailed shrew is little more than mouse-sized itself, the glands from a single specimen will provide enough venomous material to kill 200 mice. Mice injected show classic signs of envenomation: general depression, increased urination, partial paralysis, cyanosis, protruding eyes, labored breathing, and finally convulsions just before death. The venom is also apparently proteolytic—it aids in breaking down or digesting protein material and undoubtedly helps the shrew with its blast furnace-like metabolism utilize food more quickly than if its digestion had to start from scratch. The shrews are all highly active animals and specimens have been known to eat twice their own weight within 24 hours. Although this was probably something like a record, it has been reported that some species at least will starve to death in seven hours. Anything, any device or substance that will enable such an animal to utilize food more quickly and more efficiently is bound to have survival value.

At present there is no real indication that any mammals other than these few insectivores have a venomous bite. In a way that is interesting since the insectivores are very primitive mammals. There is the temptation to compare the primitive venomous mammals with elapid snakes, but beyond being an interesting exercise there is little to be gained by the effort. There are pharmacological similarities between the two venoms—both being essentially neurotoxic. But the snake's glands are modified parotid glands, not submaxillaries. Also, the elapids have hollow fangs in the *upper* jaws and the insectivores have channeled or concave incisors in the *lower* jaw. In fact, there are more similarities between the envenomating systems of the insectivores and the Helodermatidae, the venomous lizards, than with the elapid snakes. It is true that the mammals rose up from the reptiles and that at one time many more mammals may have been venomous than are now, but there is little reason to believe the venom gland system found in the insectivores today is derived from an ancestor common, as well, to the elapid snakes. There is an understandable temptation to think that way but very little reason to do so.

The insectivores probably developed their envenomating systems primarily for food-getting. Their venoms are strong

A shrew of the genus *Sorex*. There are over 40 species in this genus and some of them, at least, have a mildly venomous bite. It is deadly to their regular prey but harmless to man. It is so mild, in fact, that its existence was only recently discovered and we still do not know how many species have this capability.

enough to immobilize if not kill the prey they normally seek. They are fast-acting neurotoxins and it is often reported that the shrews try to bite their prey on the back of the head, the area closest to the central nervous system.

This concludes our discussion of the venomous mammals. Why, of all the diverse mammals found on this planet, only these few have venom is not known. If there are more, that, too, is not known. We have a great deal to learn. There may be many surprises yet awaiting us. Understandably we have focused the attention of our laboratories on venoms of most immediate concern to us—those of the snakes, some fish, and the arachnids. Those venoms can maim and kill us. The venoms of the mammals are not only obscure but also harmless to man and so they have had to wait. There can be little doubt, though, that once we understand mammalian venoms and envenomating systems better than we do now we will have a better understanding of all venoms. That, from all indications, is a goal worth striving for.

The Haitian solenodon (*Solenodon paradoxus*), a venomous insectivore only recently understood to have a toxic bite.

BIBLIOGRAPHY

Abalos, J. W. "Scorpions of Argentina," in *Venomous and Poisonous Animals and Noxious Plants of the Pacific Region,* ed. Hugh L. Keegan and W. V. Macfarlane. New York: Pergamon Press, 1963.

————, and I. Pirosky. "Venomous Argentine Serpents: Ophidism and Snake Antivenin," *ibid.*

————, and E. C. Baez. "The Spider Genus *Latrodectus* in Santiago, Del Estero, Argentina," in *Aminal Toxins,* ed. Findlay E. Russell and Paul R. Saunders. New York: Pergamon Press, 1967.

Abbott, R. Tucker. "Venom Apparatus and Geographical Distribution of *Conus gloriamaris," Notalae Naturae,* No. 40, May 3, 1967.

Ahuja, M. L., and Gurkirpal Singh. "Snakebite in India," in *Venoms,* ed. Eleanor E. Buckley and Nandor Porges. Washington, D.C.: Am. Assoc. Advanc. Science, 1956.

Allam, M. W., *et al.* "Comparison of Cortisone and Antivenin in the Treatment of Crotaline," in *Venoms,* ed. Eleanor E. Buckley and Nandor Porges. Washington, D.C.: Am. Assoc. Advanc. Science, 1956.

Allington, Herman V., and R. Raymond Allington. "Insect Bites," *Journal of the American Medical Association,* 155, No. 3, May 15, 1954.

Amaral, Afranio Do. "Snake Venoms and Their Antidotes," in *Practice of Pediatrics,* Vol. 1, ed. Brennemann-Kelly, Chap. 19N., 1963.

Ambrose, Michael S. "Snakebite in Central America," in *Venoms,* Eleanor E. Buckley and Nandor Porges. Washington, D.C.: Am. Assoc. Advanc. Science, 1956.

Amorin, Maocyr DeF. "Intermediate Nephron Nephrosis in Human and Experimental Crotalis Poisoning," in *Venomous Animals and Their Venoms,* Vol. II, ed. Wolfgang Bücherl and Eleanor E. Buckley. New York: Academic Press, 1971.

Andrews, Carl E., *et al.* "Venomous Snakebite in Florida," *Journal of the Florida Medical Association,* 55, No. 4, Apr. 1968.

Arno, Stephen F. "Interpreting the Rattlesnake," *National Parks Magazine,* 43, No. 267, Dec. 1969.

Asano, Motokazu, and Massao Itoh, "Salivary Poison of a Marine Gastropod *Neptunea arthritica bernardi,* and the Seasonal Variation of Its Toxicity in Biochemical and Pharmaceutical Compounds Derived from Marine Organisms," *Annals of the New York Academy of Sciences,* Vol. 90, Art. 3, Nov. 1960.

Ashe, James. "Some Facts about Snake Fangs," *Africana,* 3, No. 4 Dec. 1967.

Atz, James W. "Stonefishes: The World's Most Venomous Fishes," *Animal Kingdom,* 63, No. 5, 1960.

————, "The Flamboyant Zebra Fish," *ibid.,* 65, No. 2, 1962.

Aymar, Brandt (ed.). *Treasury of Snake Lore.* Greenberg, N.Y.: 1956.

Bachmayer, H., *et al.* "Chemistry of Cytotoxic Substances in Amphibian Toxins," in *Animal Toxins*, ed. F. F. Russell and P. R. Saunders. New York: Pergamon Press, 1967.

Baerg, William J., *The Tarantula*. Lawrence, Kan.: University of Kansas Press, 1958.

———, *The Black Widow and Five Other Venomous Spiders in the United States*. University of Arkansas Agricultural Experimental Station Bulletin 608. Fayetteville, Ark., Feb. 1959.

———, *Scorpions: Biology and Effect of Their Venom*. University of Arkansas Agricultural Experimental Station Bulletin 649. Fayetteville, Ark., Aug. 1961.

Ballering, R. B., *et al.* "Octopus Envenomation Through a Plastic Bag via a Salivary Proboscis," *Toxicon*, 10, No. 3, May 1972.

Balozet, Lucien. "Scorpion Venoms and Antiscorpion Serum," in *Venoms*, ed. Eleanor E. Buckley and Nandor Porges. Washington, D.C.: Am. Assoc. Advanc. Science, 1956.

———. "Scorpionism in the Old World," in *Venomous Animals and Their Venoms*, Vol. III, ed. Wolfgang Bücherl and Eleanor E. Buckley. New York: Academic Press, 1971.

Barme, Michel. "Venomous Sea Snakes of Viet Nam and Their Venoms," in *Venomous and Poisonous Animals and Noxious Plants of the Pacific Region*, ed. Hugh L. Keegan and W. V. Macfarlane. New York: Pergamon Press, 1963.

———. "Venomous Sea Snakes," in *Venomous Animals and Their Venoms*, Vol. I, ed. Wolfgang Bücherl *et al.* New York: Academic Press, 1968.

Barnes, J. H. "Studies on Three Venomous Cubomedusae," in *The Cnidaria and Their Evolution*, ed. W. J. Rees. London: Zoological Society of London, Academic Press, 1966.

———. "Extraction of Cnidarian Venom From Living Tentacle," in *Animal Toxins*, ed. F. F. Russell and P. R. Saunders. New York: Pergamon Press, 1967.

Bartsch, Paul. *Mollusks*, Vol. 10 (Series). Washington, D.C.: Smithsonian Institution, 1934.

Becak, Willy. "Karyotypes, Sex Chromosomes, and Chromosomal Evolution in Snakes," in *Venomous Animals and Their Venoms*, Vol. I, ed. Wolfgang Bücherl *et al.* New York: Academic Press, 1968.

Belluomini, Helio Emerson. "Extraction and Quantities of Venom Obtained from Some Brazilian Snakes," in *Venomous Animals and Their Venoms*, Vol. I, ed. Wolfgang Bücherl *et al.* New York: Academic Press, 1968.

Benton, Allen W., *et al.* "Bioassay and Standardization of Venom of the Honey Bee," *Nature*, 198, No. 4877, Apr. 20, 1963.

———. "Venom Collection from Honey Bees," *Science*, 142, No. 3589, Oct. 1963.

———, and R. L. Patton. "A Qualitative Analysis of the Proteins in the Venom of Honey Bees," *Journal of Insect Physiology*, Vol. 11, 1965.

———, and R. A. Morse. "Collection of the Liquid Fraction of Bee Venom," *Nature*, 210, No. 5036, May 1966.

———. "Esteras and Phosphotases of Honeybee Venom," *Journal of Apicultural Research*, 6, No. 2, 1967.

———, *et al.* "Bee Venom Tolerance in White Mice in Relation to Diet," *Science*, 145, No. 3639, Sept. 1969.

Benyajati, Chanyo, and Chaloem Puranananda, "The Evolution of Treatment of Snakebite in Thailand," in *Venomous and Poisonous Animals and Noxious Plants of the Pacific Region*, ed. Hugh L. Keegan and W. V. Macfarlane. New York: Pergamon Press, 1963.

Best, E. A. "Occurrence of the Round Stingray *Urolophus halleri cooper* in Humboldt Bay, California," *California Fish and Game*, 47, No. 4, Oct. 1961.

"Beware the Brown Recluse!" *Changing Times*, Apr. 1969.

Bishop, Sherman C. *Handbook of Salamanders*. Ithaca, N.Y.: Comstock Publishing Assoc., 1943.

Blackwelder, Richard E. *Classification of the Animal Kingdom*. Carbondale, Ill.: Southern Illinois University Press, 1963.

"Black Widow Spiders May Be Dangerous," *Home Safety Review*, 15, No. 2, Feb. 1958.

Blanquet, Richard. "A Toxic Protein from the Nematocysts of the Scyphozoan Medusa *Chrysaora quinquecirrha*," *Toxicon*, 10, No. 2, Mar. 1972.

Bogen, Emil. "The Treatment of Spider Bite Poisoning," in *Venoms*, ed. Eleanor E. Buckley and Nandor Porges. Washington, D.C.: Am. Assoc. Advanc. Science, 1956.

Bogert, Charles M. "Dentitional Phenomena in Cobras and Other Elapids with Notes on Adaptive Modification of Fangs," *Bulletin of the American Museum of Natural History*, Vol. 81, Art. 3, Apr. 1943.

———. "Snakes That Spit Their Venom," *Animal Kingdom*, 57, No. 3, June 1954.

———, and R. M. Del Campo. "The Gila Monster and Its Allies: The Relationships, Habits, and Behavior of the Lizards of the Family Helodermatidae," *Bulletin of the Museum of Natural History*, Vol. 109, Art. 1, 1956.

Bonilla, Carlos A., *et al.* "Comparative Biochemistry of *Sistrurus miliarius barbouri* and *Sistrurus catenatus tergeminus* Venoms," in *Venomous Animals and Their Venoms*, Vol. II, ed. Wolfgang Bücherl and Eleanor E. Buckley. New York: Academic Press, 1971.

Bonta, I. L., *et al.* "Method for Study of Snake Venom Induced Hemorrhages," *Toxicon*, 8, No. 1, May 1970.

Boquet, P., *et al.* "Studies on Some Antigenic Proteins and Polypeptides from *Naja nigricollis* Venom," in *Animal Toxins*, ed. F. E. Russell and P. R. Saunders. New York: Pergamon Press, 1967.

———. "Chemistry and Biochemistry of the Snake Venoms of Europe and the Mediterranean Regions," in *Venomous Animals and Their Venoms*, Vol. I, ed. Wolfgang Bücherl *et al.* New York: Academic Press, 1968.

———. "Pharmacology and Toxicology of Snake Venoms of Europe and the Mediterranean Regions," *ibid.*

Boys, Floyd, and H. M. Smith. *Poisonous Amphibians and Reptiles: Recognition and Bite Treatment.* Springfield, Ill.: Charles C Thomas, 1959.

Brand, J. M., *et al.* "Fire Ant Venoms: Comparative Analysis of Alkaloidal Components," *Toxicon*, 10, No. 3, May 1972.

Brattstrom, Bayard H. "The Fossil Pit-Vipers (Reptilia: Crotalidae) of North America," *Trans. San Diego Society of Natural History*, 12, No. 3, July 1954.

———. "Evolution of the Pit-Vipers," *ibid.*, 13, No. 11, May 1964.

Broadley, Donald G. "The Venomous Snakes of Central and South Africa," in *Venomous Animals and Their Venoms*, Vol. I, ed. Wolfgang Bücherl *et al.* New York: Academic Press, 1968.

Brodie, Edmund D., Jr. "Investigations on the Skin Toxin of the Adult Rough-Skinned Newt, *Taricha granulosa*," *Copeia*, No. 2, June 1968.

———. "Investigations of the Skin Toxin of the Red-Spotted Newt, *Notophthalmus viridescens viridescens*," *American Midland Naturalist*, 80, No. 1, July 1968.

Brown, G. W., Jr. (ed.). *Desert Biology*, Vol. I. New York: Academic Press, 1968.

Bryson, Kenneth D. "The Treatment of Chronic Arthritis with a Combination of Cobra Venom, Formic Acid, and Silicic Acid," *American Surgeon*, 20, No. 7, July 1954.

Bücherl, Wolfgang, *et al.* (eds.). *Venomous Animals and Their Venoms.* 3 vols. New York: Academic Press, 1968-1971.

———. "Classification, Biology and Venom Extraction of Scorpions," in *Venomous Animals and Their Venoms*, Vol. III, ed. Wolfgang Bücherl and Eleanor E. Buckley. New York: Academic Press, 1971.

———. "Venomous Chilpods or Centipedes," *ibid.*

———. "Spiders," *ibid.*

Buckley, Eleanor E., and Nandor Porges (eds.). *Venoms.* American Association for the Advancement of Science, Publication No. 44. Washington, D.C.: AAAS, 1956.

Burnett, J. W., and R. Goldner, "Effect of *Chrysaora quinquelcirrha* (Sea Nettle) Toxin on Rat Nerve and Muscle," *Toxicon*, 8, No. 2, Aug. 1970.

———. "Some Immunological Aspects of Sea Nettle Toxins," *Toxicon*, 9, No. 3, July 1971.

Burrell, H. *The Platypus.* Sydney: Angus & Robertson, 1927.

Calaby, J. H. "The Platypus (*Ornithorhynchus anatinus*) and Its Venomous Characteristics," in *Venomous Animals and Their Venoms*, Vol. I, ed. Wolfgang Bücherl *et al.* New York: Academic Press, 1971.

Cameron, Ann M., and R. Endean. "Venom Glands in Scatophagid Fish," *Toxicon*, 8, No. 2, Aug. 1971.

———. "The Axillary Glands of the Plotosid Catfish *Cnidoglanis macrocephalus*," *Toxicon*, 9, No. 4, Oct. 1971.

Caras, Roger A. *Dangerous to Man*. Philadelphia: Chilton, 1964.

———. *North American Mammals*. New York: Meredith, 1967.

Carlson, R. W., *et al.* "Some Pharmacological Properties of the Venom of the Scorpionfish *Scorpaena guttata*—I," *Toxicon*, 9, No. 4, Oct. 1971.

Carroll, Robert L. "Origin of Reptiles," in *Biology of the Reptilia*, Vol. 1, ed. Carl Gans *et al.* New York: Academic Press, 1969.

Castex, Mariano N., "Fresh Water Venomous Rays," in *Animal Toxins*, ed. F. E. Russell and P. R. Saunders. New York: Pergamon Press, 1967.

Chapman, David S. "The Symptomatology, Pathology, and Treatment of the Bites of Venomous Snakes of Central and Southern Africa," in *Venomous Animals and Their Venoms*, Vol. I, ed. Wolfgang Bücherl *et al.* New York: Academic Press, 1968.

Chaudhuri, D. K., *et al.* "Pharmacology and Toxicology of the Venoms of Asiatic Snakes," in *Venomous Animals and Their Venoms*, Vol. II, ed. Wolfgang Bücherl and Eleanor E. Buckley. New York: Academic Press, 1971.

Chen, K. K., and Alena Kovarikova. "Pharmacology and Toxicology of Toad Venom," *Journal of Pharmaceutical Sciences*, 56, No. 12, Dec. 1967.

Christensen, Poul Agerhohn. *South African Snake Venoms and Antivenoms*. Johannesburg: South African Institute of Medical Research, July 1955.

———. "The Preparation and Purification of Antivenoms,"

———. "Venom and Antivenom Potency Estimation," *ibid.*

———. "The Venoms of Central and South African Snakes," in *Venomous Animals and Their Venoms*, Vol. I, ed. Wolfgang Bücherl *et al.* New York: Academic Press, 1968.

———. "The Treatment of Snakebite," *South African Medical Journal*, Vol. 43, Oct. 1969.

———, and C. G. Anderson. "Observations on *Dendroaspis* Venoms," in *Animal Toxins*, ed. F. E. Russell and P. R. Saunders. New York: Pergamon Press, 1967.

Cleland, Sir John B., and R. V. Southcott. *Injuries to Man from Marine Invertebrates in the Australian Region*. Commonwealth of Australia, National Health and Medical Research Council, Special Report Series No. 12. Canberra, 1965.

Cloudsley-Thompson, J. L. *Spiders, Scorpions, Centipedes and Mites*. London: Pergamon, Press, 1968.

Cochran, Doris M. *Living Amphibians of the World*. Garden City, N.Y.: Doubleday, 1962.

———, and Coleman J. Goin. *Frogs of Colombia*. Smithsonian Institution Bulletin No. 288. Washington, D.C., 1970.

Cogger, Harold, G. "The Venomous Snakes of Australia and Melanesia," in *Venomous Animals and Their Venoms*, Vol. II, ed. Wolfgang Bücherl and Eleanor E. Buckley. New York: Academic Press, 1971.

Commonwealth Serum Laboratories. *Treatment of Snakebite in Australia*. Melbourne, Oct. 1961.

———. *Red-Black Spider Antivenene*. (Instruction leaflet.) 1966.

———. *Stone-Fish Antivenene*. Melbourne, July 1967.

———. *Treatment of Snakebite in Australia and New Guinea Using Antivenene*. Melbourne, Oct. 1967.

Comstock, John Henry. *The Spider Book*. Ithaca, N.Y.: Cornell University Press, 1940.

Corkill, Norman L. "Snake Poisoning in the Sudan," in *Venoms*, ed. Eleanor E. Buckley and Nandor Porges. Washington, D.C.: Am. Assoc. Advanc. Science, 1956.

Criley, B. R. "Development of a Multivalent Antivenin for the Family *Crotalidae*," in *Venoms*, ed. Eleanor E. Buckley and Nandor Porges. Washington, D. C.: Am. Assoc. Advanc. Science, 1956.

Crompton, John. *The Snake.* London: Faber & Faber, 1963.

Crone, H. D., and T. E. B. Keen. "Further Studies in the Biochemistry of the Toxins from the Sea Wasp *Chironex fleckeri*," *Toxicon*, 9, No. 2, Apr. 1971.

Cropp, Ben. "Sea Snakes," *Oceans*, 3, No. 2, Mar.-Apr. 1970.

Dack, G. M. *Food Poisoning* (3rd ed.). Chicago: University of Chicago Press, 1965.

Daly, J. W., *et al.* "Batracotoxin: The Active Principle of theColombian Arrow Poison Frog, *Phylobates bicolor*," *Journal of the American Chemical Society*, 87, No. 124, 1965.

———, and B. Witkop. "Chemistry and Pharmacology of Frog Venoms," in *Venomous Animals and Their Venoms*, Vol. II, ed. Wolfgang Bücherl and Eleanor E. Buckley. New York: Academic Press, 1971.

Davis, Harry T. *Poisonous Snakes of the Eastern United States with First Aid Guide.* Raleigh: North Carolina State Museum, n.d.

"Deadly Octopus Plague," *Science News*, Vol. 96, July 12, 1969.

Deas, Walter. "Venomous Octopus," *Sea Frontiers*, 16, No. 6, Nov.-Dec. 1970.

Dees, John. "Florida Snake Bite Data—1963," *Journal of the Florida Medical Association*, 49, No. 12, June 1963.

De Klobusitzky, D. "Animal Venoms in Therapy," in *Venomous Animals and Their Venoms*, ed. Wolfgang Bücherl and Eleanor E. Buckley. New York: Academic Press, 1971.

Del Pozo, E. C. "Mechanism of Pharmacological Actions of Scorpion Venoms," in *Venoms*, ed. Eleanor E. Buckley and Nandor Porges. Washington, D.C.: Am. Assoc. Advanc. Science, 1956.

Denson, K. W. E., and W. E. Rousseau. "Separation of the Coagulant Components of *Bothrops jararaca* Venom," *Toxicon*, 8, No. 1, May 1970.

Deoras, P. J. *Snakes: How To Know Them.* Bombay: Directorate of Publicity, Feb. 1959.

———. "A Study of Scorpions," *Probe*, 1, No. 2, Dec. 1961.

———. "Studies on Bombay Snakes: Snake Farm Venom Yield Records and Their Probable Significance," in *Venomous and Poisonous Animals and Noxious Plants of the Pacific Region*, ed. Hugh L. Keegan and W. V. Macfarlane. New York: Pergamon Press. 1963.

———. *Snakes of India.* New Delhi: National Book Trust, 1965.

———. "The Story of Some Indian Poisonous Snakes," in *Venomous Animals and Their Venoms*, Vol. II, ed. Wolfgang Bücherl and Eleanor E. Buckley. New York: Academic Press, 1971.

Department of the Navy, Bureau of Medicine and Surgery. *Poisonous Snakes of the World.* Washington, D.C.: Government Printing Office, 1965 (rev.).

De Sylva, Donald P. "Stingers, Biters—and Divers," *Sea Frontiers*, 13, No. 6, Nov.-Dec. 1967.

Deulofeu, Venancio, and E. A. Ruveda. "The Basic Constituents of Toad Venoms," in *Venomous Animals and Their Venoms*, Vol. II, ed. Wolfgang Bücherl and Eleanor E. Buckley. New York: Academic Press, 1971.

Deutsch, H. F., and C. R. Diniz. "Some Proteolytic Activities of Snake Venoms," in *Venoms*, ed. Eleanor E. Buckley and Nandor Porges. Washington, D.C.: Am. Assoc. Advanc. Science, 1956.

Devi, Anima, *et al.* "Anticoagulating Action of Cobra Venom," in *Venoms*, ed. Eleanor E. Buckley and Nandor Porges. Washington, D.C.: Am. Assoc. Advanc. Science, 1956.

———, ———, "Prothrombin-like Property of Russell's Viper Venom," *ibid.*

———. "The Protein and Nonprotein Constituents of Snake Venoms," in *Venomous Animals and Their Venoms*, Vol. I, ed. Wolfgang Bücherl *et al.* New York: Academic Press, 1968.

———. "The Chemistry, Toxicity, Biochemistry, and Pharmacology of North American Snake Venoms," *ibid.*, Vol. II, 1971.

Dicus, Donald R. "Current Status of Treatment of Poisonous Snakebite," *Postgraduate Medicine,* Mar. 1962.

Diniz, Carlos R., and J. M. Goncalves. "Some Chemical and Pharmacological Properties of Brazilian Scorpions," in *Venoms,* ed. Eleanor E. Buckley and Nandor Porges. Washington, D.C.: Am. Assoc. Advanc. Science, 1956.

————, and A. P. Corrado. "Venoms of Insects and Arachnids," in *Pharmacology and Toxicology of Naturally Occurring Toxins,* Vol. II, ed. H. Raskova. Oxford: Pergamon Press, 1971.

————. "Bradykinin Formation by Snake Venoms," in *Venomous Animals and Their Venoms,* Vol. I, ed. Wolfgang Bücherl *et al.* New York: Academic Press, 1968.

————. "Chemical and Pharmacological Properties of Tityus Venoms," *ibid.,* Vol. III, 1971.

Dobie, J. Frank. *Rattlesnakes.* Boston: Little, Brown, 1965.

Dodge, Natt N. *Poisonous Dwellers of the Desert.* Globe, Ariz.: Southwestern Monuments Ass., 1952.

Doery, Hazel M. "Purine Compounds in Snake Venoms," *Nature,* Vol. 177, Feb. 25, 1956.

————. "Additional Purine Compounds in the Venom of the Tiger Snake *(Notechis scutatus),*" *Nature,* Vol. 180, Oct. 19, 1957.

————. "The Separation and Properties of the Neurotoxins from the Venom of the Tiger Snake *Notechis scutatus scutatus,*" *Biochemical Journal,* 70, No. 4, 1958.

————, and Joan E. Pearson. "Haemolysins in Venoms of Australian Snakes," *Biochemical Journal,* Vol. 78, 1961.

Dowling, Herndon G. "Poisonous Snakes of Vietnam," *Animal Kingdom,* Apr. 1966.

Drummond, F. H. "Assay of Spider Venom and Antivenene in *Drosophila,*" *Nature,* Vol. 178, Aug. 4, 1956.

Dunson, William A. "The Sea Snakes Are Coming," *Natural History,* 80, No. 9, Nov. 1971.

Eadie, R. *The Life and Habits of the Platypus.* Melbourne: Stilwell & Stephens, 1935.

Edery, H., *et al.* "Pharmacological Activity of Oriental Hornet *(Vespa orientalis)* Venom," *Toxicon,* 10, No. 1, Jan. 1972.

Edmund, A. G. "Dentition," in *Biology of the Reptilia,* Vol. I, ed. Carl Gans *et al.* New York: Academic Press, 1969.

El Asmar, M. F., *et al.* "Fractionation of Scorpion *(Leiurus quinquestriatus h.* and *e.)* Venom," *Toxicon,* 10, No. 1, Jan. 1972.

Emery, Jerry A., and Findlay E. Russell. "Lethal and Hemorrhagic Properties of Some North American Snake Venoms," in *Venomous and Poisonous Animals and Noxious Plants of the Pacific Region,* ed. Hugh L. Keegan and W. V. Macfarlane. New York: Pergamon Press, 1963.

Endean, R., *et al.* "The Venom of the Piscivorous Gastropod *Conus Striatus,*" in *Animal Toxins,* ed. F. E. Russell and P. R. Saunders. New York: Pergamon Press, 1967.

————, and Mary Noble. "Toxic Material from the Tentacles of the Cubomedusan *Chironex fleckeri,*" *Toxicon,* 9, No. 3, July 1971.

Evans, H. Muir. *Sting-Fish and Seafarer.* London: Faber & Faber, 1943.

Evans, Howard E., and Mary Jane West Eberhard. *The Wasps.* Ann Arbor: University of Michigan Press, 1970.

Fairley, N. H., and Beryl Splott. "Venom Yelds in Australian Poisonous Snakes," *Medical Journal of Australia,* 1, No. 11, 1929.

Fänge, Ragnar. "The Salivary Gland of *Neptunea antiqua,*" in *Biochemistry and Pharmacology of Compounds Derived from Marine Organisms, Annals of the New York Academy of Sciences,* Vol. 90, Art. 3, Nov. 1960.

Favilli, Giovanni. "Occurrence of Spreading Factors and Some Properties of Hyaluronidases in Animal Parasites and Venoms," in *Venoms,* ed. Eleanor E. Buckley and Nandor Porges. Washington, D.C.: Am. Assoc. Advanc. Science, 1956.

Finlayson, M. H. "Arachnidism in South Africa," in *Venoms*, ed. Eleanor E. Buckley and Nandor Porges. Washington, D.C.:Am. Assoc. Advanc. Science, 1956.

Fischer, George A., and Jon L. Kabora. "Low Molecular Weight Toxins Isolated from Elapidae Venoms," in *Animal Toxins*, ed. F. E. Russell and P. R. Saunders. New York: Pergamon Press, 1967.

Fish, Charles J., and Mary Curtis Cobb. *Noxious Marine Animals of the Central and Western Pacific Ocean*. U.S. Dept. of the Interior, Fish and Wildlife Service, Research Report No. 36. Washington, D.C.: Government Printing Office, 1954.

Fitzsimons, D. C. *A Guide to the Durban Snake Park*. Durban, S.A., n.d.

Fitzsimons, Vivian F. M. *Snakes of Southern Africa*. Cape Town: Purnell & Sons, 1962.

Flecher, Hugo. "Injuries by Unknown Agents to Bathers in North Queensland," *Medical Journal of Australia*, Jan. 27, 1945.

————. "Fatal Stings to North Queensland Bathers," *ibid.*, Jan. 12, 1952.

————. "Irukandji Sting to North Queensland Bathers without Production of Weals but with Severe General Symptoms," *ibid.*, July 19, 1952.

————. "Injuries Produced by Marine Animals in Tropical Australia," Unpublished manuscript, prior to 1955.

Fluno, John A. "Wasps as Enemies of Man," *Bulletin of the Entomological Society of America*, 7, No. 3, Sept. 1961.

Free, John B., and Colin G. Butler. *Bumblebees*. London: Collins, 1959.

Freeman, Shirley E., and R. J. Turner. "Cardiovascular Effects of Cnidarian Toxins: A Comparison of Toxins Extracted from *Chiropsalmus quadrigatus* and *Chironex fleckeri*," *Toxicon*, 10, No. 1, Jan. 1972.

Freire-Maia, L., *et al.* "Effects of Purified Scorpion Toxin on Respiratory Movements in the Rat," *Toxicon*, 8, No. 4, Nov. 1970.

Friese, U. Erich. "Death in a Small Package: The Blue-Ringed Octopus," *Animal Kingdom*, 75, No. 2, Apr. 1972.

Froom, Barbara. *The Massasauga Rattlesnake*. Federation of Ontario Naturalists, Special Publication No. 2. Don Mills, Ont.: n.d.

Fuhrman, Frederick A. "Tetrodotoxin," *Scientific American*, 217, No. 2, Aug. 1967.

Gans, Carl, *et al.* (eds.). *Biology of the Reptilia* (3 vols.). New York: Academic Press, 1969-70.

Ganthavorn, S. "A Case of Cobra Bite," *Toxicon*, 9, No. 3, July 1971.

Gennaro, Joseph F. Jr. "Studies on Snake Venom Toxicity." Address to the Thirty-fifth Annual Meeting of the Florida State Veterinary Medicine Association, Oct. 6, 1959.

————. "Observations on the Treatment of Snakebite in North America," in *Venomous and Poisonous Animals and Noxious Plants of the Pacific Region*, ed. Hugh L. Keegan and W. V. Macfarlane. New York: Pergamon Press, 1963.

Gertsch, Willis J. *American Spiders*. Princeton, N.J.: Van Nostrand, 1949.

Ghiretti, F. "Toxicity of Octopus Saliva against Crustacea," *Annals of the New York Academy of Sciences*, Vol. 9, Art. 3, Nov. 1960.

Ghosh, B. N., and N. K. Sarkar. "Active Principles of Snake Venoms," in *Venoms*, ed. Eleanor E. Buckley and Nandor Porges. Washington, D.C.: Am. Assoc. Advanc. Science, 1956.

————, and D. K. Chaudhuri. "Chemistry and Biochemistry of the Venoms of Asiatic Snakes," in *Venomous Animals and Their Venoms*, Vol. I, ed. Wolfgang Bücherl *et al.* New York: Academic Press, 1968.

Gilbo, Catherine M., and N. W. Coles. "An Investigation of Certain Components of the Venom of the Female Sydney Funnel-Web Spider, *Atrax robustus* Cambr.," *Australian Journal of Biological Science*, Vol. 17, 1964.

Gilluly, Richard H. "Consequences of a Sea-Level Canal: A New Study of Sea Snakes Reinforces Ecologists' Concerns about the Proposed Canal," *Science News*, Vol. 99, Jan. 16, 1971.

Gingrich, W. C., and J. C. Hohenodel. "Standardization of Polyvalent Antivenin," in *Venoms,* ed. Eleanor E. Buckley and Nandor Porges. Washington, D.C.: Am. Assoc. Advanc. Science, 1956.

Gitter, S., Chaya Moroz, *et al. Pathogenesis of Snake Venom Intoxication.* (Part I,
Neurotoxins; Part II, Coagulation Factors and Hemorrhagins; Part III,
Hemolysins). Beilinson Hospital Supply, No. 10. Tel Aviv, 1961.

————, and A. De Vries. "Symptomatology, Pathology, and Treatment of Bites
by Near Eastern, European, and North African Snakes," in *Venomous
Animals and Their Venoms,* Vol. I, ed. Wolfgang Bücherl *et al.* New York:
Academic Press, 1968.

Gladney, William J. *Controlling the Brown Recluse Spider.* U.S. Dept. Of Agriculture, Leaflet No. 556. Washington, D.C.: Government Printing Office,
1972.

Glenn, W. G., *et al.* "Quantitative In Vitro and In Vivo Studies of Rattlesnake
Venom *(Crotalus atrox),* in *Venomous and Poisonous Animals and Noxious
Plants of the Pacific Region,* ed. Hugh L. Keegan and W. V. Macfarlane.
New York: Pergamon Press, 1963.

Gowanloch, James Nelson, and Clair A. Brown. *Poisonous Snakes, Plants and
Black Widow Spider of Louisiana.* New Orleans: La. Dept. of Conserva-
tion, 1943.

Griffiths, Mervyn. *Echidnas.* London: Pergamon Press, 1968.

Grocott, Robert G., and Glendy G. Sadler. *The Poisonous Snakes of Panama.*
Mount Hope, Canal Zone, Oct. 1958.

Gruber, Michael. "Lancers of the Reef," *Sea Frontiers,* 17, No. 1, Jan.-Feb. 1971.

Haast, William E., and Melvin L. Winer. "Complete and Spontaneous Recovery
from the Bite of a Blue Krait *(Bungarus caeruleus)," American Journal of
Tropical Medicine and Hygiene,* 4, No. 6, Nov. 1955.

Haberman, E. "Chemistry, Pharmacology, and Toxicology of Bee, Wasp, and
Hornet Venoms," in *Venomous Animals and Their Venoms,* Vol. III, ed.
Wolfgang Bücherl and Eleanor E. Buckley. New York: Academic Press,
1971.

Habermehl, Gerhard. "Toxicology, Pharmacology, Chemistry, and Biochemis-
try of Salamander Venom," in *Venomous Animals and Their Venoms,* Vol.
II ed. Wolfgang Bücherl and Eleanor E. Buckley. New York: Academic
Press, 1971.

Hadidian, Z. "Proteolytic Activity and Physiologic and Pharmacologic Actions
of *Agkistrodon piscivorus* Venom," in *Venoms,* ed. Eleanor E. Buckley and
Nandor Porges. Washington, D.C.: Am. Assoc. Advanc. Science, 1956.

Halstead, Bruce W. "Animal Phyla Known to Contain Poisonous Marine Animals," in *Venoms,* ed. Eleanor E. Buckley and Nandor Porges. Washington, D.C.: Am. Assoc. Advanc. Science, 1956.

————. "Weever Stings and Their Medical Treatment," *U.S. Armed Forces Medical Journal,* 8, No. 10, Oct. 1957.

————. *Dangerous Marine Animals.* Cambridge, Md.: Cornell Maritime Press,
1959.

————, and Lois A. Mitchell. "A Review of the Venomous Fishes of the Pacific
Area," in *Venomous and Poisonous Animals and Noxious Plants of the Pacific
Region,* ed. Hugh L. Keegan and W. V. Macfarlane. New York: Perga-
mon Press, 1963.

————, and Arthur E. Dalgleish. "The Venom Apparatus of the European
Star-Gazer *Uranoscopus scober* Linnaeus," in *Animal Toxins,* ed. F. E. Russell and P. R. Saunders. New York: Pergamon Press, 1967.

————, and D. A. Courville. *Poisonous and Venomous Marine Animals of the World*
(3 vols.). Washington, D.C.: Government Printing Office, 1965-70.

————, and D. D. Danielson. "Death from the Depths," *Oceans Magazine,* 3, No.
6, Nov-Dec. 1970.

————. "Venomous Fish," in *Venomous Animals and Their Venoms,* Vol. II, ed.
Wolfgang Bücherl and Eleanor E. Buckley. New York: Academic Press,
1971.

————. "Venomous Coelenterates: Hydroids, Jellyfishes, Corals, and Sea
Anemones," *ibid.,* Vol. III.

Harmon, R. W., and C. B. Pollard. *Bibliography of Animal Venoms.* Gainesville: University of Florida Press, 1948.

Hartman, William J., *et al.* "Pharmacologocally Active Amines and Their Biogenesis in the Octopus," in *Annals of the New York Academy of Sciences,* Vol. 90, Art. 3, 1960.

Hawkey, Christine, and C. Symons. "Coagulation of Primate Blood by Russell's Viper Venom," *Nature,* 210, No. 5032, Apr. 9, 1966.

Henriques, S. B., and Olga B. Henriques. "Pharmacology and Toxicology of Snake Venoms," in *Pharmacology and Toxicology of Naturally Occurring Toxins,* Vol. I, ed. H. Raskova. Oxford: Pergamon Press, 1971.

Hills, Ralph G., and Warfield M. Firor. "The Use of More Potent Venom for Intractable Pain," *American Surgeon,* 18, No. 9, Sept. 1952.

Hoge, Alphonse Richard, *et al.* "Neotropical Pit Vipers, Sea Snakes, and Coral Snakes," in *Venomous Animals and Their Venoms,* Vol. II, ed. Wolfgang Bücherl and Eleanor E. Buckley. New York: Academic Press, 1971.

Horen, W. Peter. "Arachnidism in the United States," *Modern Medicine,* Dec. 9, 1963.

Horsfall, William R. *Medical Entomology: Arthropods and Human Disease.* New York: Ronald Press, 1962.

Hoyt, Murray. *The World of Bees.* New York: Bonanza Books, 1965.

Irwin, R. S. "Funnel-Web Spider Bite," *Medical Journal of Australia,* Dec. 6, 1952.

Irwin, Richard L., *et al.* "Toxicity of Elapidae Venoms and an Observation in Relation to Geographical Location," *Toxicon,* 8, No. 1, May 1970.

Jackman, A. I. "Cobra Venom Therapy in the Neuroses," *Diseases of the Nervous System,* 15, No. 4, Apr. 1954.

Jacobs, Werner. "Floaters of the Sea," *Natural History,* 71, No. 7, Aug.-Sept. 1962.

Jaeger, Robert G. "Toxic Reaction to Skin Secretions of the Frog *Phrynomerus bifasciatus,*" *Copeia,* No. 1, 1971.

Jaques, R. "The Hyaluronidase Content of Animal Venoms," in *Venoms,* ed. Eleanor E. Buckley and Nandor Porges. Washington, D.C.: Am. Assoc. Advanc. Science, 1956.

Jensen, H., and U. Westphal. "Chemical Structure and Interrelationships of Toad Poisons," in *Venoms,* ed. Eleanor E. Buckley and Nandor Porges. Washington, D.C.: Am. Assoc. Advanc. Science, 1956.

Jimenez-Porras, Jesus M. "Differentiation between *Bothrops nummifer* and *Bothrops picadoi* by Means of the Biochemical Properties of Their Venoms," in *Animal Toxins,* ed. F. E. Russell and P. R. Saunders. New York: Pergamon Press, 1967.

———. "Biochemistry of Snake Venoms," in *Snake Venoms and Envenomation,* ed. Sherman A. Minton, Jr. New York: Marcel Dekker, 1971.

Jobin, F., and M. P. Esnouf. "Coagulant Activity of Tiger Snake *(Notechis scutatus)* Venoms," *Nature,* 211, No. 5051, Aug. 20, 1966.

Johnson, Bob Duel. "Some Interrelationships of Selected *Crotalus* and *Agkistrodon* Venom Properties and Their Relative Lethalities." Unpublished dissertation, Arizona State University, 1967.

Kaire, G. H. "Observations of Some Funnel-Web Spider *(Atrax* sp.) and Their Venoms with Particular Reference to *Atrax robustus,*" *Medical Journal of Australia,* Vol. 2, Aug. 24, 1963.

———. "The Sydney Funnel-Web Spider *(Atrax robustus)* in Captivity," *Victoria Naturalist,* Vol. 81, June 1964.

Kaiser, Erich. "Enzymatic Activity of Spider Venoms," in *Venoms,* ed. Eleanor E. Buckley and Nandor Porges. Washington, D.C.: Am. Assoc. Advanc. Science, 1956.

———, and Robert Kramar. "Biochemistry of the Cytotoxic Action of Amphibian Poisons," in *Animal Toxins,* ed. F. E. Russell and P. R. Saunders. New York: Pergamon Press, 1967.

———, and H. Michl. "Chemistry and Pharmacology of the Venoms of *Bothrops* and *Lachesis,*" in *Venomous Animals and Their Venoms,* Vol. II, ed. Wolfgang Bücherl and Eleanor E. Buckley. New York: Academic Press, 1971.

Karlsson, E., and D. Eaker. "Isolation of the Principal Neurotoxins of *Naja naja* Subspecies from the Asian Mainland," *Toxicon*, 10, No. 3, May 1972.

Kaston, B. J. *The Black Widow Spider*. New England Museum of Natural History, Museum Leaflet No. 5. Boston, 1937.

Katz, N. L., and C. Edwards. "The Effect of Scorpion Venom on the Neuromuscular Junction of the Frog," *Toxicon*, 10, No. 2, Mar. 1972.

Keegan, Hugh L. "Antivenins Available for Treatment of Envenomation by Poisonous Snakes, Scorpions, and Spiders," in *Venoms*, ed. Eleanor E. Buckley and Nandor Porges. Washington, D.C.: Am. Assoc. Advanc. Science, 1956.

————. *Some Venomous and Noxious Animals of The Far East*. Contrib. Dept. Entomology, Medical General Lab (406), U.S. Army Medical Comm. Japan, 1960.

————. "Caterpillars and Moths as Public Health Problems," in *Venomous and Poisonous Animals and Noxious Plants of the Pacific Region*, ed. Hugh L. Keegan and W. V. Macfarlane. New York: Pergamon Press, 1963.

————. "Centipedes and Millipedes as Pests in Tropical Areas," *ibid*.

————, and W. V. Macfarlane (eds.) *Venomous and Poisonous Animals and Noxious Plants of the Pacific Region*. New York: Pergamon Press, 1963.

————. "Venomous Spiders of the Pacific Area," in *ibid*.

Keele, C. A. "Venoms and the Causes of Pain," *New Scientist*, No. 327, Feb. 21, 1963.

Keen, T. E. B. "Comparison of Tentacle Extracts from *Chiropsalmus quadrigatus* and *Chironex Fleckeri*," *Toxicon*, 9, No. 3, July 1971.

Klauber, Laurence M. *A Statistical Study of the Rattlesnakes*. San Diego Society of Natural History, Occasional Paper No. 5, Aug. 30, 1939.

————. *Rattlesnakes: Their Habits, Life Histories, and Influence on Mankind* (2 vols.). Berkeley: University of California Press, 1956.

————. "Some Factors Affecting the Gravity of Rattlesnake Bite," in *Venoms*, ed. Eleanor E. Buckley and Nandor Porges. Washington, D.C.: Am. Assoc. Advanc. Science, 1956.

————. "Classification, Distribution, and Biology of the Venomous Snakes of Northern Mexico, the United States, and Canada: *Crotalus* and *Sistrurus*," in *Venomous Animals and Their Venoms*, Vol. II, ed. Wolfgang Bücherl and Eleanor E. Buckley. New York: Academic Press, 1971.

Klein, Warren E. "Portuguese Man-of-War Sting," *U.S. Armed Forces Medical Journal*, 11, No. 8.

Klemmer, Konrad. "Methods of Classification of Venomous Snakes," in *Venomous Animals and Their Venoms*, Vol. I, ed. Wolfgang Bücherl *et al*. New York: Academic Press, 1968.

————. "Classification and Distribution of European, North African, and North and West Asiatic Snakes," *ibid*.

Kloske, W., and James J. Summary. "The Coagulation of Human Plasma by *Trimeresurus okinavensis* Venom," *American Journal of Medical Sciences*, May 1965.

Kocholaty, W. F., *et al*. "Toxicity and Some Enzymatic Properties and Activities in the Venoms of Crotalidae, Elapidae, and Viperidae," *Toxicon*, 9, No. 2, Apr. 1971.

Kochva, Elazar, and Carl Gans. "The Structure of the Venom Gland and Secretion of Venom in Viperid Snakes," in *Animal Toxins*, ed. F. E. Russell and P. R. Saunders. New York: Pergamon Press, 1967.

————. "Salivary Glands of Snakes," in *Snake Venoms and Envenomation*, ed. Sherman A. Minton, Jr. New York: Marcel Dekker, 1971.

Kohn, Alan J. "Recent Cases of Human Injury Due to Venomous Marine Snails of the Genus *Conus*," *Hawaii Medical Journal*, Vol. 17, July-Aug. 1958.

————, *et al*. "Preliminary Studies on the Venom of the Marine Snail *Conus*," in *Annals of the New York Academy of Sciences*, Vol. 90, Art. 3, Nov. 1960.

————. "Venomous Marine Snails of the Genus *Conus*," in *Venomous and Poisonous Animals and Noxious Plants of the Pacific Region*, ed. Hugh L. Keegan and W. V. Macfarlane. New York: Pergamon Press, 1963.

Koszalka, M. F. "A Case of Hemoglobinuria in Bee Sting," *Bulletin U.S. Army Medical Dept.*, 9, No. 212, Mar. 1949.

Kulkarni, M. E., and S. S. Rao. "Antigenic Composition of the Venoms of Poisonous Snakes of India," in *Venoms,* ed. Eleanor E. Buckley and Nandor Porges. Washington, D.C.: Am. Assoc. Advanc. Science, 1956.

Lane, Charles E. "The Toxin of *Physalia* Nematocysts," *Annals of the New York Academy of Sciences,* Vol. 90, Art. 3, Nov. 17, 1960.
————. "The Deadly Fisher," *National Geographic,* 123, No. 3, Mar. 1963.
————. "Recent Observations on the Pharmacology of *Physalia* Toxin," in *Animal Toxins,* ed. F. E. Russell and P. R. Saunders. New York: Pergamon Press, 1967.
Lane, Frank W. *Kingdom of the Octopus.* New York: Sheridan House, 1960.
Lanham, Url. *The Insects.* New York: Columbia University Press, 1964.
Larsen, J. B., and Charles E. Lane. "Direct Action of *Physalia* Toxin on Frog Nerve and Muscle," *Toxicon,* 8, No. 1, May 1970.
Leake, Chauncey D. "Development of Knowledge about Venoms," in *Venoms,* ed. Eleanor E. Buckley and Nandor Porges. Washington, D.C.: Am. Assoc. Advanc. Science, 1956.
Lee, Chen-Yuan, *et al.* "Cholinesterase Inactivating Activity of Snake Venom," in *Venoms,* ed. Eleanor E. Buckley and Nandor Porges. Washington, D.C.: Am. Assoc. Advanc. Science, 1956.
————, ————. "Mode of Neuromuscular Blocking Action of the Desert Black Snake Venom," *Toxicon,* 9, No. 4, Oct. 1971.
————. "Elapid Neurotoxins and Their Mode of Action," in *Snake Venoms and Envenomation,* ed. Sherman A. Minton, Jr. New York: Marcel Dekker, 1971.
Leviton, Alan E. "The Venomous Terrestrial Snakes of East Asia, India, Malaya, and Indonesia," in *Venomous Animals and Their Venoms,* Vol. I, ed. Wolfgang Bücherl *et al.* New York: Academic Press, 1968.
Licht, Lawrence E., and Bobbi Low. "Cardiac Response of Snakes after Ingestion of Toad Parotoid Venom," *Copeia,* No. 3, 1968.
Linaweaver, Paul G. "Toxic Marine Life," *U.S. Navy Medical News Letter,* 51, No. 5, Mar. 8, 1968.
Lockhard, William E. "Treatment of Snakebite," *Journal of the American Medical Association,* 193, No. 5, Aug. 2, 1965.
Loeb, L., *et al. The Venom of Heloderma.* Carnegie Institute Publication No. 177. Washington, D.C., 1913.
Loosanoff, Victor L. *Jellyfishes and Related Animals.* U.S. Dept. of the Interior, Fishery Leaflet No. 535, Feb. 1962.
Lowe, Charles H. "Appraisal of Research on Fauna of Desert Environments," in *Deserts of the World,* ed. William G. McGinnies *et al.* Tucson: University of Arizona Press, 1968.
Lumpkin, William R., and Warfield M. Firor. "Evaluation of the Bryson Treatment of Arthritis," *American Surgeon,* 20, No. 7, July 1954.
Luther, Wolfgang. "Distribution, Biology, and Classification of Salamanders," in *Venomous Animals and Their Venoms,* Vol. II, ed. Wolfgang Bücherl and Eleanor E. Buckley. New York: Academic Press, 1971.
Lutz, Bertha. "Venomous Toads and Frogs," in *Venomous Animals and Their Venoms,* Vol. II, ed. Wolfgang Bücherl and Eleanor E. Buckley. New York: Academic Press, 1971.

Macfarlane, R. G. "Russell's Viper Venom, 1934-64," *British Journal of Haematology,* Vol. 13, 1967.
Macht, David I. "New Developments in the Pharmacology and Therapeutics of Cobra Venom," Transactions of the American Therapeutic Society, Vol. 40, 1940.
Madon, Minoo B., and Ronald E. Hall. "First Record of *Loxosceles rufescens* (Dufour) in California," *Toxicon,* 8, No. 1, May 1970.
Mahanama, Gamini. *Snake-Bite Myths and Cures.* Loris, Dec. 1967.
Malette, W. G., *et al.* "The Pathophysiologic Effects of Rattlesnake Venom (*Crotalus atrox*)," in *Venomous and Poisonous Animals and Noxious Plants of the Pacific Region,* ed. Hugh L. Keegan and W. V. Macfarlane. New York: Pergamon Press, 1963.

Maretic, Zvonimir. "Venom of an East African Orthognath Spider," in *Animal Toxins*, ed. F. E. Russell and P. R. Saunders. New York: Pergamon Press, 1967.

———. "Latrodectism in Mediterranean Countries, Including South Russia, Israel, and North Africa," in *Venomous Animals and Their Venoms*, Vol. III, ed. Wolfgang Bücherl and Eleanor E. Buckley. New York: Academic Press, 1971.

Märki, F., and B. Witkop. "The Venom of the Colombian Arrow Poison Frog *Phyllobates bicolor*," *Separatum Experientia*, Vol. 19, 1963.

Marr, A. G. M., and E. H. Baxter. "Effect of Proteolytic Enzymes on the Venom of the Sea Wasp *Chironex fleckeri*," *Toxicon*, 9, No. 4, Oct. 1971.

Marsh, J. A. *Cone Shells of the World*. Brisbane: Jacaranda Press, 1964.

Marsh, Helene. "Preliminary Studies of the Venoms of Some Vermivorous Conidae," *Toxicon*, 8, No. 4, Nov. 1970.

———. "The Caseinase Activity of Some Vermivorous Cone Shell Venoms," *Toxicon*, 9, No. 1, Jan. 1971.

Maschwitz, Ulrich W. J., and Werner Kloft. "Morphology and Function of the Venom Apparatus of Insects—Bees, Wasps, Ants, and Caterpillars," in *Venomous Animals and Their Venoms*, Vol. III, ed. Wolfgang Bücherl and Eleanor E. Buckley. New York: Academic Press, 1971.

Master, R. W. P., and S. Srinivasa Rao. "Identification of Enzymes and Toxins in Venoms of India Cobra and Russell's Viper after Starch Gel Electophoresis," *Journal of Biological Chemistry*, 236, No. 7, July 1961.

Mazzotti, Luis, and M. A. Bravo-Becherelle. "Scorpionism in the Mexican Republic," in *Venomous and Poisonous Animals and Noxious Plants of the Pacific Region*, ed. Hugh L. Keegan and W. V. Macfarlane. New York: Pergamon Press, 1963.

McCollough, Newton C. "Venomous Snake Bite," *Journal of the Florida Medical Association*, 49, No. 12, June 1963.

———, and Joseph F. Gennaro. "Coral Snake Bites in the United States," *ibid.*

———, ———. "Evaluation of Venomous Snake Bite in the Southern United States," *ibid.*

———, ———. "Summary of Snake Bite Treatment," *ibid.*

———, "Emergency Room Treatment of Venomous Snakebite," *Journal of the Florida Medical Association*, 55, No. 4, Apr. 1968.

———, and Joseph F. Gennaro. "Diagnosis, Symptoms, Treatment, and Sequelae of Envenomation by *Crotalus adamanteus* and Genus *Ancistrodon*," *ibid.*

———, ———. "Treatment of Venomous Snakebite in the United States," in *Snake Venoms and Envenomation*, ed. Sherman A. Minton, Jr. New York: Marcel Dekker, 1971.

McCrone, John D., and Robert J. Hatala. "Isolation and Characterization of Lethal Component from the Venom of *Latrodectus mactans mactans*," in *Animal Toxins*, ed. F. E. Russell and P. R. Saunders. New York: Pergamon Press, 1967.

McGinnies, William G., *et al.* (eds.). *Deserts of the World*. Tucson: University of Arizona Press, 1968.

McIntosh, Max E., and Dead D. Watt. "Biochemical-Immunochemical Aspects of the Venom from the Scorpion *Centruroids sculpturatus*, in *Animal Toxins*, ed. F. E. Russell and P. R. Saunders. New York: Pergamon Press, 1967.

McMichael, Donald F. "Dangerous Marine Molluscs," Supplementary Bulletin, Post Graduate Committee in Medicine, Proc. First Internat. Convention of Life Saving Techniques, 18, No. 12, Mar. 1963.

———. "Slides of Dangerous Molluscs," *ibid.*

———. "Classification, Distribution, Venom Apparatus and Venoms, Symptomatology of Stings (Venomous Mollusks)," in *Venomous Animals and Their Venoms*, Vol. III, ed. Wolfgang Bücherl and Eleanor E. Buckley. New York: Academic Press, 1971.

Meadows, Paul E., and Findlay E. Russell. "Milking of Arthropods," *Toxicon*, 8, No. 4, Nov. 1970.

Merck, Sharp & Dohme. *Lyovac Antivenin* (Latrodectus mactans). Direction circular, 1965.

Meyer, Kuno, and Horst Linde. "Collection of Toad Venoms and Chemistry of the Toad Venom Steroids," in *Venomous Animals and Their Venoms,* Vol. II, ed. Wolfgang Bücherl and Eleanor E. Buckley. New York: Academic Press, 1971.

Micks, Don W. "Clinical Effects of the Sting of the Puss Caterpillar *Megalopyge opercularis s.* and *a.* on Man," *Texas Reports on Biology and Medicine,* 10, No. 2, Summer 1952.

———. "Insects and Other Arthropods of Medical Importance in Texas," *ibid.,* 18, No. 4, Winter 1960.

———. "An Outbreak of Dermatitus Due to the Grain Itch Mite *Pyemotes ventricosus* Newport," *ibid.,* 20, No. 2, Summer 1962.

———. "The Current Status of Necrotic Arachnidism in Texas," in *Venomous and Poisonous Animals and Noxious Plants of the Pacific Region,* ed. Hugh L. Keegan and W. V. Macfarlane. New York: Pergamon Press, 1963.

Middelbrook, R. E., *et al.* "Isolation and Purification of Toxin from *Millepora dichotoma,*" *Toxicon,* 9, No. 4, Oct. 1971.

Miller, D. G. Jr. "Masssive Anaphylaxis from Insect Stings," in *Venoms,* ed. Eleanor E. Buckley and Nandor Porges. Washington, D.C.: Am. Assoc. Advanc. Science, 1956.

Minton, Sherman A., Jr. "Some Properties of North American Pit Viper Venoms an Their Correlation with Phylogeny," in *Venoms,* ed. Eleanor E. Buckley and Nandor Porges. Washington, D.C.: Am. Assoc. Advanc. Science, 1956.

———. "Snakebite," *Scientific American,* 196, No. 1, Jan. 1957.

———. "Observations on Toxicity and Antigenic Makeup of Venoms from Juvenile Snakes," in *Animal Toxins,* ed. F. E. Russell and P. R. Saunders. New York: Pergamon Press, 1967.

———. "Venoms of Desert Animals," in *Desert Biology,* Vol. I, ed. G. W. Brown, Jr. New York: Academic Press, 1968.

———. "The Feeding Strike of The Timber Rattlesnake," *Journal of Herpetology,* Vol. 3, 1969.

———, and Madge R. Minton. *Venomous Reptiles.* New York: Scribner's, 1969.

———. *Snake Venoms and Envenomation.* New York: Marcel Dekker, 1971.

———. "Indentification of Poisonous Snakes," *ibid.*

Mohamed, A. H., *et al.* "Effects of *Naja nigricollis* Venom on Blood and Tissue Histamine," *Toxicon,* 9, No. 2, Apr. 1971.

———. "The Effect of *Naja nigricollis* Venom on Blood Clotting," *ibid.*

Moore, Richard E., and Paul J. Scheuer. "Palytonin: A New Marine Toxin from a Coelenterate," *Science,* Vol. 172, Apr. 30, 1971.

Morales, Flavio, *et al.* "Effect of Several Agents on the Lethal Action of Two Common Venoms," in *Venomous and Poisonous Animals and Noxious Plants of the Pacific Region,* ed. Hugh L. Keegan and W. V. Macfarlane. New York: Pergamon Press, 1963.

Moreton, Ann. "Spiders—Feared, Revered, and Loathed the World Over," *Smithsonian Magazine,* 2, No. 5, Aug. 1971.

Morgan, F. G. "The Australian Taipan *Oxyuranus scutellatus scutellatus* (Petus)," in *Venoms,* ed. Eleanor Buckley and Nandor Porges. Washington, D.C.: Am. Assoc. Advanc. Science, 1956.

Moroz, Chaya, *et al.* "Neurotoxic Protein of *Vipera palestinae* Venom," in *Animal Toxins,* ed. F. E. Russell and P. R. Saunders. New York: Pergamon Press, 1967.

Morris, Ramona, and Desmond Morris. *Men and Snakes.* New York: McGraw-Hill, 1965.

Morse, Roger A., and Allen W. Benton. "Notes on Venom Collection from Honeybees," *Bee World,* 45, No. 4, 1964.

Mowbray, Beverly D. "The Deadly Sea Snail," *Frontiers,* Philadelphia Academy of Science, Feb. 1965.

Mueller, Harry Louis, "Further Experiences with Severe Allergic Reactions to Insect Stings," *New England Journal of Medicine,* 261, No. 8, Aug. 20, 1959.

——— "Serious Allergic Reactions to Insect Stings," *American Journal of Nursing,* 60, No. 8, Aug. 1960.

Munjal, D., and W. B. Elliott. "Studies of Antigenic Fractions in Honey-Bee (*Apis mellifera*) Venom," *Toxicon,* 9, No. 3, July 1971.

National Safety Council. *Black-Widow Spiders* (rev.) Data Sheet 258, n.d.

New York Zoological Society. *Emergency Snakebite Procedures.* Reptile Dept. Leaflet No. 9. Rev., May 16, 1966.

North, Edgar A., and Hazel M. Doery. "Antagonism between the Actions of Staphylococcal Toxin and Tiger Snake Venom," *Nature,* Vol. 181, May 31, 1958.

O'Connor, Rod, *et al.* "The Venom of the Honeybee (*Apis mellifera*): I, General Character," in *Animal Toxins,* ed. F. E. Russell and P. R. Saunders. New York: Pergamon Press, 1967.

"Octopus Venom Isolated," *Science News,* Vol. 93, June 29, 1968.

Oliver, James A. *The Prevention and Treatment of Snakebite.* New York: N.Y. Zoological Society, 1952.

————. *The Natural History of North American Amphibians and Reptiles.* New York: Van Nostrand, 1955.

O'Rourke, Fergus J. "The Toxicity of Black Widow Spider Venom," in *Venoms,* ed. Eleanor E. Buckley and Nandor Porges. Washington, D.C.: Am. Assoc. Advanc. Science, 1956.

Ouyang, Chaoho, and Shoei-Yn Shiau. "Relationship between Pharmacological Actions and Enzymatic Activities of the Venom of *Trimeresurus gramineus,*" *Toxicon,* 8, No. 2, Aug. 1970.

Parrish, Henry M. "Early Excision and Suction of Snakebite Wounds in Dogs," in *Venoms,* ed. Eleanor E. Buckley and Nandor Porges. Washington, D.C.: Am. Assoc. Advanc. Science, 1956.

————, *et al.* "Human Allergy Resulting from North American Snake Venoms," *Journal of the Florida Medical Association,* Vol. 43, May 1957.

————. "Mortality from Snakebites, United States, 1950-1954," *Public Health Report,* Vol. 72, Nov. 1957.

————. "Deaths from Bites and Stings of Venomous Animals and Insects in the United States," *A.M.A. Archives of International Medicine,* Vol. 104, Aug. 1959.

————. "Treatment of Poisonous Snakebites: Present Status of Incision and Excision," *Journal of the Indiana State Medical Association,* 53, No. 10, Oct. 1960.

————. "Analysis of 460 Fatalities from Venomous Animals in the United States," *American Journal of Medical Sciences,* 245, No. 2, Feb. 1963.

————. "Intravenous Antivenin in Clinical Snake Venom Poisoning," *Missouri Medicine,* 60, No. 3, Mar. 1963.

————. "Poisonous Snakebites in North Carolina," *North Carolina Medical Journal,* 25, No. 3, Mar. 1964.

————. "Texas Snakebite Statistics," *Texas State Journal of Medicine,* Vol. 60, July 1964.

————. "Pit Viper Bites in Pennsylvania," *Pennsylvania Medical Journal,* Vol. 67, Aug. 1964.

————. "Characteristics of Snakebites in Missouri," *Missouri Medicine,* Oct. 1964.

————, *et al.* "Counting California's Snakebites," *California Medicine,* Vol. 101, Nov. 1964.

————, *et al.* Snakebite: A Pediatric Problem," *Clinical Pediatrics,* 4, No. 4, Apr. 1965.

————. "Comments on Snakebites in New Jersey," *Journal of the Medical Society of New Jersey,* 62, No. 6, June 1965.

————. "Comparison between Snakebites in Children and Adults," *Pediatrics,* 36, No. 2, Aug. 1965.

————. "Nature of Poisonous Snakebites in New York," *New York State Journal of Medicine,* 65, No. 17, Sept. 1, 1965.

————, and M. S. Khan. "Bites by Foreign Venomous Snakes in the United States," *American Journal of Medical Sciences,* 251, No. 2, Feb. 1966.

————, *et al.* "Poisonous Snakebites Causing No Envenomation," *Postgraduate Medicine,* 39, No. 3, Mar. 1966.

————. "Incidence of Treated Snakebites in the United States," *Public Health Reports,* 81, No. 3, Mar. 1966.

————. "Bites by Cottonmouths (*Ancistrodon piscivorus*) in the United States," *Southern Medical Journal,* 60, No. 4, Apr. 1967.

————, and M. S. Khan. "Bites by Coral Snakes: Reports of 11 Representative Cases," *American Journal of Medical Sciences,* 253, No. 5, May 1967.

————. "Pitfalls in Treating Pit Viper Bites," *Medical Times,* 95, No. 8, Aug. 1967.

————, and Carole A. Carr. "Bites by Copperheads (*Ancistrodon contortrix*) in the United States," *Journal of the American Medical Association,* Vol. 201, Sept. 18, 1967.

————, and William B. Neser. "Epidemicological Methods in Studying Venomous Snakebites," in *Animal Toxins,* ed. F. E. Russell and P. R. Saunders. New York: Pergamon Press, 1967.

————. "Current Concepts of Snakebite Treatment," *U.S. Navy Medical News Letter,* Dec. 20, 1968.

————, and Robert A. Hayes. "Hospital Management of Pit Viper Venenations," in *Snake Venoms and Envenomation,* ed. Sherman A. Minton, Jr. New York: Marcel Dekker, 1971.

Patton, Walter Scott. "Insects, Ticks, Mites, and Venomous Animals of Medical and Veterinary Importance," Parts I and II, *Medical and Public Health. Croyden: H. R. Grubb, Ltd., 1929, 1931.*

Pearson, Oliver P. "A Toxic Substance from the Salivary Glands of a Mammal (Short-tailed Shrew)," in *Venoms,* ed. Eleanor E. Buckley and Nandor Porges. Washington, D.C.: Am. Assoc. Advanc. Science, 1956.

Pesce, Hugo, and Alvaro Delgado. "Poisoning from Adult Moths and Caterpillars," in *Venomous Animals and Their Venoms,* Vol. III, ed. Wolfgang Bücherl and Eleanor E. Buckley. New York: Academic Press, 1971.

Peters, James A. *Dictionary of Herpetology.* New York: Hafner Pub., 1964.

Picarelli, Zuleika P., and J. R. Do Valle. "Pharmacological Studies on Caterpillar Venoms," in *Venomous Animals and Their Venoms,* Vol. III, ed. Wolfgang Bücherl and Eleanor E. Buckley. New York: Academic Press, 1971.

Picken, L. E. R., and R. J. Skaer. "A Review of Researches on Nematocysts," in *Cnidaria and Their Evolution,* ed. W. J. Rees. London: Zoological Society of London, Academic Press, 1966.

Pickwell, Gayle. *Amphibians and Reptiles of the Pacific States.* Palo Alto, Calif.: Stanford University Press, 1947.

Pinney, Roy. "Poisonous Snake-Bite Therapy." Unpublished manuscript, n.d.

Pirosky, I., and J. W. Abalos. "Spiders of the Genus *Latrodectus* in Argentina. Latrodectism and *Latrodectus* Antivienin," in *Venomous and Poisonous Animals and Noxious Plants of the Pacific Region,* ed. Hugh L. Keegan and W. V. Macfarlane. New York: Pergamon Press, 1963.

Pollard, C. B. "Venom Research: A Challenge to the Various Sciences," in *Venoms,* ed. Eleanor E. Buckley and Nandor Porges. Washington, D.C.: Am. Assoc. Advanc. Science, 1956.

Pope, Clifford H. *The Reptile World: A Natural History of the Snakes, Lizards, Turtles, and Crocodilians.* New York: Knopf, 1966.

Pournelle, George H. "Classification, Biology, and Description of the Venom Apparatus of Insectivores of the Genera *Solenodon, Neomys,* and *Blarina,*" in *Venomous Animals and Their Venoms,* Vol. I, ed. Wolfgang Bücherl *et al.* New York: Academic Press, 1968.

Pucek, Michalina. "Chemistry and Pharmacology of Insectivore Venoms," in *Venomous Animals and Their Venoms,* Vol. I, ed. Wolfgang Bücherl *et al.* New York: Academic Press, 1968.

Puranananda, Choloem. "Treatment of Snakebite Cases in Bangkok," in *Venoms,* ed. Eleanor E. Buckley and Nandor Porges. Washington, D.C.: Am. Assoc. Advanc. Science, 1956.

Ramsey, H. W., *et al.* "Fractionation of Coral Snake Venom. Preliminary Studies on the Separation and Characterization of the Protein Fractions," *Toxicon,* 10, No. 1, Jan. 1972.

Randel, H. W., and C. B. Doan. "Caterpillar Urticaria in the Panama Canal Zone: Report of Five Cases," in *Venoms*, ed. Eleanor E. Buckley and Nandor Porges. Washington, D.C.: Am. Assoc. Advanc. Science, 1956.

Rao, Shanta S., and S. S. Rao. "Proteases of Cobra (*Naja naja*) Venom," in *Venoms*, ed. Eleanor E. Buckley and Nandor Porges. Washington, D.C.: Am. Assoc. Advanc. Science, 1956.

Raskova, H. (ed.). *Pharmacology and Toxicology of Naturally Occurring Toxins* (2 vols.) Section 71: *International Encyclopedia of Pharmacology and Therapeutics*. Oxford: Pergamon Press, 1971.

Ray, Sammy M., and D. V. Aldrich. "Ecological Interactions of Toxic Dinoflagellates and Molluscs in the Gulf of Mexico," in *Animal Toxins*, ed. F. E. Russell and P. R. Saunders. New York: Pergamon Press, 1967.

Rees, W. J. (ed.). *The Cnidaria and Their Evolution*. London: Zoological Society of London, Academic Press, 1966.

Reid, H. Alistair. "Three Fatal Cases of Sea Snakebite," in *Venoms*, ed. Eleanor E. Buckley and Nandor Porges. Washington, D.C.: Am. Assoc. Advanc. Science, 1956.

————. "Prolonged Coagulation Defect (Defibrination Syndrome) in Malayan Viper Bite," *The Lancet*, Mar. 23, 1963.

————. "Clinical Effects of Bites by Malayan Viper (*Agkistrodon rhodostoma*)," *ibid.*

————. "Snakebite in Malaya," in *Venomous and Poisonous Animals and Noxious Plants of the Pacific Region*, ed. Hugh L. Keegan and W. V. Macfarlane. New York: Pergamon Press, 1963.

————. "Defibrination by *Agkistrodon rhodostoma* Venom," in *Animal Toxins*, ed. F. E. Russell and P. R. Saunders. New York: Pergamon Press, 1967.

————. "Symptomatology, Pathology, and Treatment of Land Snake Bite in India and Southeast Asia," in *Venomous Animals and Their Venoms*, Vol. 1, ed. Wolfgang Bücherl *et al.* New York: Academic Press, 1968.

————. "Snakebite in the Tropics," *U.S. Navy Medical News Letter*, 53, No. 2, Feb. 1969.

————. "The Principles of Snakebite Treatment," in *Snake Venoms and Envenomation*, ed. Sherman A. Minton, Jr. New York: Marcel Dekker, 1971.

Rhoten, William, and Joseph F. Gennaro, Jr. "Treatment of the Bite of a Mojave Rattlesnake," *Journal of the Florida Medical Association*, 55, No. 4, Apr. 1968.

Rittenbury, Max S., and Lawrence D. Hanback. "Snake Antivenin," *Archives of Surgery*, Vol. 99, Aug. 1966.

Robertson, Phyllis L. "A Morphological and Functional Study of the Venom Apparatus in Representatives of Some Major Groups of Hymenoptera," *Australian Journal of Zoology*, 16, No. 1, Feb. 1968.

Romer, J. D. "Notes on the White-lipped Pit Viper *Trimeresurus albolabris* Gray in Hong Kong," in *Venomous and Poisonous Animals and Noxious Plants of the Pacific Region*, ed. Hugh L. Keegan and W. V. Macfarlane. New York: Pergamon Press, 1968.

Rosenfeld, G., *et al.* "Coagulant Proteolytic and Hemolytic Properties of Some Snake Venoms," in *Venomous Animals and Their Venoms*, Vol. I, ed. Wolfgang Bücherl *et al.* New York: Academic Press, 1968.

————. "Symptomatology, Pathology, and Treatment of Snake Bites in South America," in *ibid.*, Vol. II, 1971.

Rotberg, A. "Lepidopterism in Brazil," in *Venomous Animals and Their Venoms*, Vol. III, ed. Wolfgang Bücherl and Eleanor E. Buckley. New York: Academic Press, 1971.

Rothschild, Lord. *A Classification of Living Animals*. London: Longmans, 1965.

Russell, Findlay E., and R. D. Lewis. "Evaluation of the Current Status of Therapy for Stingray Injuries," in *Venoms*, ed. Eleanor E. Buckley and Nandor Porges. Washington, D.C.: Am. Assoc. Advanc. Science, 1956.

————, and A. van Harreveld. "Cardiovascular Effects of the Venoms of the Round Stingray *Urobatis halleri*," *ibid.*

————. *Stingray Injuries*. Public Health Reports, U.S. Dept. of Health, Education and Welfare, Vol. 74, No. 10, Oct. 1959.

————, and Jerry A. Emery. "Venom of the Weevers *Trachinus draco* and *Trachinus vipera*," *Annals of the New York Academy of Sciences*, Vol. 90, Art. 3, Nov. 1960.

————. "Muscle Relaxants in Black Widow Spider (*Latrodectus mactans*) Poisoning," *American Journal of Medical Science*, 243, No. 2, Feb. 1962.

————, and R. S. Scharffenberg. *Bibliography of Snake Venoms and Venomous Snakes*. West Covina: Bibliography Assoc., 1964.

————. "Toxic Marine Animals." Excerpts from a paper presented to the First Inter-American Naval Research Conference, San Juan, July 1965.

————, and L. Lauritzen. "Antivenins," *Transcripts of the Royal Society of Tropical Medicine and Hygiene*, 60, No. 6, 1966.

————, and Paul R. Saunders (eds.). *Animal Toxins*. New York: Pergamon Press, 1967.

————, *et al.* "Clinical Use of Antivenin Prepared from Goat Serum," *Toxicon*, 8, No. 1, May 1970.

————. "Bite by the Spider *Phidippus formosus:* Case History," *Toxicon*, 8, No. 2, Aug. 1970.

————. "Pharmacology of Toxins of Marine Organisms," in *Pharmacology and Toxicology of Naturally Occuring Toxins*, Vol. ed. H. Raskova. Oxford: Pergamon Press, 1971.

————, and Harold W. Puffer. "Pharmacology of Snake Venoms," in *Snake Venoms and Envenomation*, ed. Sherman A. Minton, Jr. New York: Marcel Dekker, 1971.

Russel, N.J. "Snake Venom," *Animal Kingdom*, Aug. 1968.

San Martin, Pablo Rubens. "The Venomous Ants of the Genus *Solenopsis*," in *Venomous Animals and Their Venoms*, Vol. III, ed. Wolfgang Bücherl and Eleanor E. Buckley. New York: Academic Press, 1971.

Sarkar, N.K., and Anima Devi. "Enzymes in Snake Venoms," in *Venomous Animals and Their Venoms*, Vol. I, ed. Wolfgang Bücherl *et al.* New York: Academic Press, 1968.

Saunders, Paul R. "Venom of the Stonefish *Synanceja verrucosa*," *Science*, 129, No. 3344, Jan. 30, 1959.

————, and Peter B. Taylor. "Venom of the Lionfish *Pterois volitans*," *American Journal of Physiology*, 197, No. 2, Aug. 1959.

————. "Venom of the Stonefish *Synanceja horrida* (Linnaeus)," *Arch* Int. *Pharmacodyn*, 123, No. 1-2, 1959.

————. "Venoms of Scorpionfishes," *Proc West Pharmacology Society*, Vol. 2, 1959.

————. "Sting by a Venomous Lionfish," *U.S. Armed Forces Medical Journal*, 11, No. 2, Feb. 1960.

————. "Pharmacological and Chemical Studies of the Venom of the Stonefish (Genus *Synanceja*) and Other Scorpionfishes," *Annals of the New York Academy of Sciences*, Vol. 90, Art. 3, Nov. 17, 1960.

————. "Purification and Properties of the Lethal Fraction of the Venom of the Stonefish *Synanceja horrida* (Linnaeus)," Acta *Biochimica et Biophysica*, Vol. 1952, 1961.

Savory, Theodore H. "Daddy Longlegs," *Scientific American*, Oct. 1962.

Schaeffer, R. C. Jr., *et al.* "Some Chemical Properties of the Venom of the Scorpionfish *Scorpaena guttata*," *Toxicon*, 9, No. 1, Jan. 1971.

Schall, Joseph. "Sea Snakes—Mysterious Mariners," *Frontiers*, 33, No. 3, Feb. 1969.

Schenberg, S., and F. A. Pereira Lima. "*Phoneutria nigriventer* Venom—Pharmacology and Biochemistry of Its Components," in *Venomous Animals and Their Venoms*, Vol. III, ed. Wolfgang Bücherl and Eleanor E. Buckley. New York: Academic Press, 1971.

Schmidt, Karl P. "Anent the 'Dangerous' Bushmaster," *Copeia*, No. 3, Aug. 26, 1957.

————, and R. F. Inger. *Living Reptiles of the World*. Garden City, N.Y.: Hanover House, 1957.

"Scorpion Stings and Heart Diseases," *Science News*, 95, No. 4, Jan. 1969.

Shannon, Frederich R. "Comments on the Treatment of Reptile Poisoning," in *Venoms,* ed. Eleanor E. Buckley and Nandor Porges. Washington, D.C.: Am. Assoc. Advanc. Science, 1956.

Shaw, Charles E. "A Pit Separates the Vipers," *Zoonooz,* March 1970.

———. "The Coral Snakes Genera *Micrurus* and *Micruroides* of the United States and Northern Mexico," in *Venomous Animals and Their Venoms,* Vol. II, ed. Wolfgang Bücherl and Eleanor E. Buckley, 1971. New York: Academic Press, 1971.

Slotta, Karl H., *et al.* "The Direct and Indirect Hemolytic Factors from Animal Venoms," in *Animal Toxins,* ed. F. E. Russell and P. R. Saunders. New York: Pergamon Press, 1967.

Smart, John, *et al. A Handbook for the Identification of Insects of Medical Importance.* London: British Museum (Natural History), 1965.

Smith, David S. "Structure of the Venom Gland of the Black Widow Spider *Latrodectus mactans.* A Preliminary Light and Electromicroscopic Study," in *Animal Toxins,* ed., F. E. Russell and P. R. Saunders. New York: Pergamon Press, 1967.

Smith, J. L. B. "Two Rapid Fatalities from Stonefish Stabs," *Copeia,* No. 3, Aug. 26, 1957.

Snakebite. Bilingual pamphlet. Bangkok: Siamese Red Cross Society, n.d.

Snow, Keith R. *The Arachnids: An Introduction.* New York: Columbia University Press, 1970.

Songdahl, J. H., and C. E. Lane. "Some Pharmacological Characteristicis of the Alphabet Cone, *Conus spurius atlanticus, Toxicon,* 8, No. 4, Nov. 1970.

Southcott, R. V. "Venomous Jellyfish," *Good Health,* Jan. 1960.

———. "Coelenterates of Medical Importance," in *Venomous and Poisonous Animals and Noxious Plants of the Pacific Region,* ed. Hugh L. Keegan and W. V. Macfarlane. New York: Pergamon Press, 1963.

———. "Injuries to Man from Marine Invertebrates in the Australian Seas," Documenta Geigy, Nautilus, Oct. 1969.

Sowder, Wilson T., and Geogre W. Gehres. "Snakebites in Florida," *Journal of the Florida Medical Association,* 49, No. 12, June 1963.

———. "Snakebite Myths and Misinformation," *ibid.,* 55, No. 4, April 1968.

Sparger, Charles Forrest. "Problems in the Management of Rattlesnake Bites," *Archives of Surgery,* Vol. 98, Jan. 1969.

Stackhouse, John. *Australia's Venomous Wildlife.* London: Paul Hamlyn, 1970.

Stahnke, Herbert L. *Scorpions* (rev. ed.). Tempe: Poisonous Animals Research Laboratory, Arizona State University, 1956.

———, and Bob D. Johnson. "*Aphonopelma* Tarantula Venom," in *Animal Toxins,* ed. F. E. Russell and P. R. Saunders. New York: Pergamon Press, 1967.

Stanic, M. "Allergenic Properties of Venom Hypersensitiveness in Man and Animals," in *Venoms,* ed. Eleanor E. Buckley and Nandor Porges. Washington, D.C.: Am. Assoc. Advanc. Science, 1956.

Steinitz, H. "Observations on *Pterois volitans* (L.) and Its Venom," *Copeia,* No. 2, 1959.

Steward, J. W. *The Tailed Amphibians of Europe.* New York: Taplinger, 1970.

Stickel, William H. *Venomous Snakes of the United States and Treatment of Their Bites.* U.S. Dept. Interior, Fish and Wildlife Service, Wildlife Leaflet No. 339. Washington, D.C., June 1952.

Stillway, L. W., and C. T. Lane. "Phospholipase in the Nematocyst Toxin of *Physalia physalis,*" *Toxicon,* 9, No. 3, July 1971.

Stimson, A. C., and H. T. Englehardt. "Snake Bite Needn't Be Fatal," *National Safety News,* May 1960.

"The Stone Fish," *African Wildlife,* 10, No. 1, March 1956.

"Stone-Fish and Other Antivenenes," *Medical Journal of Australia,* 1, No. 19, May 9, 1959.

Strydon, D. J., and D. P. Bates. "I. Preliminary Studies on the Separation of Toxins of Elapidae Venoms," *Toxicon,* 8, No. 3, Sept. 1970.

Su, C., *et al.* "Pharmacological Properties of the Neurotoxin of Cobra Venom," in *Animal Toxins,* ed. F. E. Russell and P. R. Saunders. New York: Pergamon Press, 1967.

Suarez, G., *et al.* "*Loxosceles laeta* Venom—Partial Purification," *Toxicon,* 9, No. 3, July 1971.

Sutherland, Struan. "The Ringed Octopus Bite: A Unique Medical Emergency," *Medical Journal of Australia,* Sept. 2, 1967.

———. "Toxins and Mode of Envenomation of the Common Ringed or Blue-banded Octopus," *ibid.,* May 3, 1969.

———, *et al.* "Octopus Neurotoxins: Low Molecular Weight Non-Immunologic Toxins Present in the Saliva of the Blue-ringed Octopus," *Toxicon,* 8, No. 3, Dec. 1970.

Swaroop, C. C., and B. Grab. "The Snakebite Mortality Problem in the World," in *Venoms,* ed. Eleanor E. Buckley and Nandor Porges, Washington, D.C.: Am. Assoc. Advanc. Science, 1956.

Syder, C. C., *et al.* "A Definitive Study of Snakebite," *Journal of the Florida Medical Association,* 55, No. 4, Apr. 1968.

Taboada, Oscar. *Medical Entomology.* Bethesda, Md.: National Naval Medical Center, 1967.

Thompson, T. E., and I. Bennett. "Physalia Nematocysts: Utilized by Mollusks for Defense," *Science,* Vol. 166, Dec. 19, 1969.

Tidswell, F. *Australian Venoms.* Sydney: Dept. of Public Health, 1906.

Tinkham, Ernest R. "The Deadly Nature of Gila Monster Venom," in *Venoms,* ed. Eleanor E. Buckley and Nandor Porges. Washington, D.C.: Am. Assoc. Advanc. Science, 1956.

———. "Bite Symptoms of the Red-legged Widow Spider (*Latrodectus bishopi*)," *ibid.*

———. "The Biology of the Gila Monster," in *Venomous Animals and Their Venoms,* Vol. II, ed. Wolfgang Bücherl and Eleanor E. Buckley. New York: Academic Press, 1971.

———. "The Venom of the Gila Monster," *ibid.*

Tokuyama, T., *et al.* "The Structure of Batrachotoxinin A, a Novel Steroidal Alkaloid from the Colombian Arrow Poison Frog, *Phyllobates aurotaenia,*" *Journal of the American Chemical Society,* 90, No. 1917, 1968.

Trethewie, E. R. "Pharmacological Effects of Australian Snake Venoms," in *Venoms,* ed. Eleanor E. Buckley and Nandor Porges. Washington, D.C.: Am. Assoc. Advanc. Science, 1956.

———. "The Pharmacology and Toxicology of the Venoms of the Snakes of Australia and Oceania," in *Venomous Animals and Their Venoms,* Vol, II, ed. Wolfgang Bücherl and Eleanor E. Buckley. New York: Academic Press, 1971.

———. "The Pathology, Symptomatology, and Treatment of Snake Bite in Australia," *ibid.*

———. "Detection of Snake Venom in Tissue," in *Snake Venoms and Envenomation,* ed. Sherman A. Minton, Jr. New York: Marcel Dekker, 1971.

Trinca, G. F. "The Treatment of Snakebite," *Medical Journal of Australia,* Feb. 23, 1963.

Trinca, John C., and Peter Schiff. "Deadly Sea Wasp," *Sea Frontiers,* 16, No. 1, Jan.-Feb. 1970.

Truitt, J. O. *A Guide to the Snakes of South Florida.* Miami: Hurricane House, 1968.

Tu, Anthony T., *et al.* "Chemical Differences in the Venoms of Genetically Different Snakes," in *Animal Toxins,* ed. F. E. Russell and P. R. Saunders. New York: Pergamon Press, 1967.

Tu, Tsuchih. "Toxicological Studies on the Venom of the Sea Snake *Laticauda laticaudata affinis,*" in *Animal Toxins,* ed. F. E. Russell and P. R. Saunders. New York: Pergamon Press, 1967.

Tweedie, M. W. F. *Poisonous Animals of Malaya.* Singapore: Malaya Pub. House, 1941.

Tyler, Albert. "An Auto-Antivenin in the Gila Monster and Its Relation to a Concept of Natural Auto-Antibodies," in *Venoms,* ed. Eleanor E. Buckley and Nandor Porges. Washington, D.C.: Am. Assoc. Advanc. Science, 1956.

Van Der Walt, S. J., and F. J. Joubert. "Studies on Puff Adder (*Bitis arietans*) Venom," *Toxicon,* 9, No. 2, Apr. 1971.

"Venomous Fish Stings," *Medical News Letter,* Vol. 34, No. 1.

Vesey-Fitzgerald, Brian. *The Worlds of Ants, Bees, and Wasps.* London, Eng.: Pelham Books, 1969.

Vick, James A., *et al.* "Pathophysiological Studies of Ten Snake Venoms," in *Animal Toxins,* ed. F. E. Russell and P. R. Saunders. New York: Pergamon Press, 1967.

———, and James Lipp. "Effect of Cobra and Rattlesnake Venoms on the Central Nervous System of the Primate," *Toxicon,* 8, No. 1, May 1970.

Viosca, Percy Jr. *Poisonous Snakes of Louisiana.* New Orleans: Louisiana Wild Life and Fisheries Comm., n.d.

Visser, John. *Poisonous Snakes of Southern Africa.* Cape Town: Howard Timmins, 1966.

Vogt, W. "What Is Toxin?" *Toxicon,* 8, No. 3, Sept. 1970.

Webster, Dwight A. "Toxicity of the Spotted Newt, *Notophthalmus viridescens,* to Trout," *Copeia,* No. 1, Mar. 25, 1960.

Weis, R., and R. J. McIsaac. "Cardiovascular and Muscular Effects of Venom from Coral Snake, *Micrurur fulvius,*" *Toxicon,* 9, No. 3, July 1971.

Werler, John E. *Poisonous Snakes of Texas.* Texas Fish and Game Comm., Bulletin No. 31. Austin, July 1963 (rev.).

———, and Hugh L. Keegan. "Venomous Snakes of the Pacific Area," in *Venomous and Poisonous Animals and Noxious Plants of the Pacific Region,* ed., Hugh L. Keegan and W. V. Macfarlane. New York: Pergamon Press, 1963.

Wheeler, William Morton. *Ants: Their Structure, Development, and Behavior.* New York: Columbia University Press, 1910, 1965.

Wheeling, C. H., and Hugh L. Keegan. "Effects of Scorpion Venom on the Tarantula," *Toxicon,* 10, No. 3, May 1972.

Whitley, G. P. *Poisonous and Harmful Fishes.* Commonwealth of Australia Council on Scientific and Industrial Research, Bulletin No. 159. Melbourne, 1943.

Whittaker, V. P. "Pharmacologically Active Choline Esters in Marine Gastropods," *Annals of the New York Academy of Sciences,* Vol. 90, Art. 3, Nov. 1960.

Whittemore, F. W., and Hugh L. Keegan. "Medically Important Scorpions in the Pacific Area," in *Venomous and Poisonous Animals and Noxious Plants of the Pacific Region,* ed. Hugh L. Keegan and W. V. Macfarlane. New York: Pergamon Press, 1963.

Wiener, Saul. "The Australian Red Back Spider (*Latrodectus hasseltii*): I. Preparation of Antiserum by the Use of Venom Absorbed on Aluminum Phosphate," *Medical Journal of Australia,* May 5, 1956.

———. "The Australian Red Back Spider (*Latrodectus hasseltii*): II. Effect of Temperature on the Toxicity of Venom," *ibid.,* Sept. 1, 1956.

———. "The Sydney Funnel-Web Spider (*Atrax robustus*): I. Collection of Venom and Its Toxicity in Animals," *ibid.,* Sept. 14, 1957.

———. "Stone-Fish Sting and Its Treatment," *ibid.,* Aug. 16, 1958.

———. "Observations on the Venom of the Stone Fish (*Synanceja trachynis*)," *ibid.,* 1, No. 19, May 1, 1959.

———. "The Sydney Funnel-Web Spider (*Atrax robustus*): II. Venom Yield and Other Characteristics of the Spider in Captivity," *ibid.,* Nov. 7, 1959.

———. "The Production and Assay of Stone-Fish Antivenene," *ibid.,* Nov. 14, 1959.

———. "Venoms Yields and Toxicity of the Venoms of Male and Female Tiger Snakes," *ibid.,* Nov. 5, 1960.

———. "Active Immunization of Man Against the Venom of the Australian Tiger Snake (*Notechis scutatus*)," *American Journal of Tropical Medicine and Hygiene,* 9, No. 3, May 1960.

———. "The Sydney Funnel-Web Spider (*Atrax robustus*): III. The Neutralization of Venom by Haemolymph," *Medical Journal of Australia,* Mar. 25, 1961.

———. "Antigenic and Electrophoretic Properties of Funnel-Web Spider (*Atrax robustus*) Venom," in *Venomous and Poisonous Animals and Noxious Plants of the Pacific Region,* ed. Hugh L. Keegan and W. V. Macfarlane. New York: Pergamon Press, 1963.

Wilson, Edward O., and Fred E. Regnier, Jr. "The Evolution of the Alarm-Defense System in the Formicine Ants," *American Naturalist,* 105, No. 943, May-June 1971.

Witkop, Bernhard. "Poisonous Animals and Their Venoms," *Journal of the Washington Academy of Science,* Vol. 55, Mar. 1965.

Wittle, L. W., *et al.* "Isolation and Partial Purification of a Toxin from *Millepora alcicornis,*" *Toxicon,* 9, No. 4, Nov. 1971.

Wolfson, Fay H. "Cones," *Animal Kingdom,* June 1970.

———. "Look Don't Touch: Cones, Part II," *Animal Kingdom,* Aug. 1970.

Wood, J. T., *et al.* "Treatment of Snake Venom Poisoning with ACTH and Cortisone," *Virginia Medical Monthly,* Mar. 1955.

Worrell, Eric. *Reptiles of Australia.* Sydney: Angus & Robertson, 1963.

———. *Dangerous Snakes of Australia and New Guinea* (5th ed.). Sydney: Angus & Robertson, 1963.

Yarom, Rena, and Karl Braun. "Cardiovascular Effects of Scorpion Venom Morphological Changes in the Myocardium," *Toxicon,* 8, No. 1, May 1970.

Zaki, Omer, *et al.* "Black Mamba Venom and Its Fractions," *Archives of Pathology,* Vol. 89. Jan. 1970.

Zlotkin, E., *et al.* "A New Toxic Protein in the Venom of the Scorpion *Androctonus Australis* Hector," *Toxicon,* 9, No. 1, Jan. 1971.

———. "The Effect of Scorpion Venom on Blowfly Larvae—A New Method for the Evaluation of Scorpion Venom's Potency," *ibid.*

———. "Proteins in Scorpion Venoms Toxic to Mammals and Insects," *Toxicon,* 10, No. 3, May 1972.

INDEX

S